LIFE
ON THE
EDGE

AMAZING CREATURES
THRIVING IN EXTREME
ENVIRONMENTS

LIFE
ON THE
EDGE

AMAZING CREATURES
THRIVING IN EXTREME
ENVIRONMENTS

MICHAEL GROSS

WITH A NEW AFTERWORD
BY THE AUTHOR

PERSEUS PUBLISHING
Cambridge, Massachusetts

Many of the designations used by manufacturers and sellers to distinguish their products are claimed as trademarks. Where those designations appear in this book and Perseus Publishing was aware of a trademark claim, the designations have been printed in initial capital letters.

Copyright © 1998 by Plenum Press, New York.
Afterword to the paperback edition © 2001 by Michael Gross

Cataloging-in-Publication Data is available from the Library of Congress
ISBN 0-7382-0445-5

Perseus Publishing is a member of the Perseus Books Group.

Find us on the World Wide Web at http://www.perseuspublishing.com

Perseus Publishing books are available at special discounts for bulk purchases in the U.S. by corporations, institutions, and other organizations. For more information, please contact the Special Markets Department at HarperCollins Publishers, 10 East 53rd Street, New York, NY 10022, or call 1-212-207-7528.

First paperback printing, January 2001

1 2 3 4 5 6 7 8 9 10—03 02 01

Contents

Of Extremists and Eccentrics
A Personal Preface

> Only connect! . . . Only connect the prose and the passion, and
> the two will be exalted . . .
> E. M. Forster, *Howards End*

Eccentrics live longer, happier, and healthier lives than conformist normal citizens, according to the neuropsychologist David Weeks. In 1984, Weeks started the first large-scale study of human beings who might be described as "splenetic," but have no psychiatric condition. His findings, which he summarized together with the journalist Jamie James in a coffee-table book, *Eccentrics* (1995), reveal that our society would be poorer without its eccentrics. Conformists may be easier to reign over, but they have neither written *Ulysses* nor developed the relativity theory. And not very many serious politicians ever managed to become nearly as popular as Joshua Abraham Norton, the King of California and Emperor of the USA, sadly overlooked by most history books. Asked whether he was eccentric himself, Weeks replied: "I'd be extremely proud to be thought of as eccentric, but I don't think I've done enough yet to deserve it."

Some other inhabitants of our planet, however, deserve to be called thus, according to James Lovelock, the creator of the controversial Gaia theory and self-confessed nonconformist. He once compared those microbes that prefer extremely hostile conditions—and that are the heroes of this book—to the eccentrics of human society. If there is a link between eccentricity and the preference for extreme conditions, it may have been my hidden eccentric disposition, which directed me toward the extremes when I crossed the lawn that separates the chemistry and life sciences departments on the Regensburg campus, looking for a topic for my final-year thesis. As a chemistry student, I had come to like the ways and means of physical chemistry, but I wasn't particularly keen to study the dynamics of benzene–water mixtures or hypercritical mercury vapors. The influence of my then-girlfriend and lab partner—a collaboration that has shifted to the field of bringing up children—had supplied me with a rather informal knowledge in biochemistry, a subject that some German chemistry departments are still trying to ignore. The extracurricular readings in modern molecular biology served as an antidote to the history-heavy chemistry courses. *Only connect!*—applying physical chemistry to biomolecules, that was an attractive option. In my supervisor, Prof. Rainer Jaenicke, I found a kindred spirit, an exiled physical chemist in the Biology Department, who set me to work on a sufficiently eccentric project: the molecular adaptation of microorganisms to high hydrostatic pressures such as those found in the deep sea.

To generate high pressure in the laboratory, you need several kilograms worth of heavy metal equipment, which may have scared off faint-hearted biologists. Thus, the high-pressure research in the Jaenicke group was nearly always carried out by a small number of chemists who had found their way across the lawn. The technical challenges are also to be blamed for the fact that adaptation to high pressures is by far not as intensively studied as the high-temperature response. Hence, before trying to find out how the adapted microbes cope with the pressure 10 kilometers below sea level, the more basic question to address is why normal, non-adapted organisms can be killed by one-quarter of that pressure. While my wife-to-be pressurized innocent little gut bacteria to search for their stress proteins, I picked one particular cell component suspected of limiting the pressure resistance of bacteria, namely the ribosome, the cell's protein factory.

The subject, which grew in various directions over the course of four years, earned me my first degree, the doctorate, and a modest reputation within the small community of high-pressure researchers. In the spring of 1993, however, I decided to give up this ecological niche and venture out into the big wide world, specifically to Oxford—I was awarded a European Molecular Biology Organization (EMBO) fellowship to study protein folding by analyzing the structural tendencies of protein fragments (peptides) in the group of Sheena Radford at the Oxford Centre for Molecular Sciences. No more extreme conditions, and no more complex systems, thought I. However, when I took up the fellowship in May, my new supervisor had different plans for me. She asked me to investigate how the molecular chaperone GroEL interacts with its substrate proteins. GroEL, which we will meet again in Chapter 3, is a double ring molecule with 14 subunits, as well as being a so-called heat shock protein, so both the extreme conditions and the complex molecular systems came back to me.

In the same year I suffered a severe outbreak of writing addiction—a condition that had been dormant inside me since the faraway times when I used to invest more time in my school's magazine than in homework. It began quite harmlessly with the idea to gain work experience in science journalism after finishing my thesis. This required publishable sample articles, which were indeed published and set me on th track of writing regularly for a major newspaper (*Süddeutsche Zeitung*) and for the German edition of *Scientific American* (*Spektrum der Wissenschaft*). The work experience never happened due to the lack of spare time between Regensburg and Oxford, but the habit of writing for one or two hours after finishing lab and family duties stuck. As the list of my journalistic publications grew, I realized there was a common motif in many of the contributions. Most of them deal with complex molecular systems such as can be found in the cell and are also aimed at modern submicrometer-scale technologies. This idea led to my first book, *Expeditionen in den Nanokosmos*, a juxtaposition of biological and technological nanoscale systems, which was published in October 1995.

For the follow-up, I did not have to search long for a topic. At some stage it must have occurred to me that I had been dealing with extreme conditions and stress all my scientific life, first the high-pressure biochemistry, then the function of a heat shock protein, and that those people who happen not to work in the same field tend to be (rightly) amazed to hear that organisms can thrive under thousandfold atmospheric pressure, in boiling water, or in saturated salt brine. Thus, as soon as I had finished work on the *Nanokosmos* I was able to put together a viable concept for *Life on the Edge*, which, again, evolved in partial overlap with my regular newspaper and magazine contributions.

The book starts with an introduction to life, its basic requirements and limits, leading to an attempt at defining the "normality" from which life's eccentrics deviate. Then we embark on an armchair voyage around the extremely hostile places on our planet and make the acquaintance of their well-adapted inhabitants. Their niches can be quite small, such as caves underneath the ground, or really vast, such as the ice fields of Antarctica. We will face abrupt temperature drops, fatal pressure changes, and various chemical stress factors. And everywhere we will encounter organisms that not only tolerate these stress conditions, but even cannot live without them. After these impressions you may be curious to know how these organisms manage to cope with all these hostile conditions. Chapter 3 will give you some of the many different answers that nature has come up with. Due to the importance of biochemical details for the stress responses and adaptation mechanisms, you may find this chapter a little bit more demanding than the bulk of the text. Don't give up, because afterward it will be downhill all the way. The amazing abilities of various microbes to deal with extreme conditions have also excited the imagination of both biotechnologists and medical researchers, and applications affecting our everyday lives may not be that far off. Once upon a time, conditions were extreme virtually everywhere on our planet. At the time when life originated, the Earth was an entirely different place from what it is today. Therefore, the origin and early evolution of life can very well be counted as "life on the edge" and is discussed in detail in Chapter 5. At the very end we will raise our views beyond the horizon of our home planet and ask whether there is life out there, in the extreme environments that may exist on other planets.

To prevent the stress pervading the pages of this book from inflicting the reader as well, I have scattered stress-relieving Sidelines and biographical Profiles of some relevant scientists all over it. (Of course there are more relevant scientists working in these fields than I can present in detail; some of them are mentioned in the text, and most of them should turn up in the list of references. My apologies for any involuntary omissions.) A third kind of diversion, labeled Focus, is to be handled with care. These are meant to provide stress-resistant and studious readers with more detailed accounts of current molecular biology issues related to the main topic. Less ambitious readers can safely skip these, as the understanding of subsequent sections does not depend on them in any way.

Of course, a book not only arises from night shifts at the computer; it also owes a great debt to conversations, correspondence, and e-mail exchange with many kind and inspiring people, including family members of three generations, friends, colleagues, and editors. Many thanks to all who gave hints and tips, commented on parts of the German or the English manuscript, or just asked a question reminding me that science can be really fascinating, even for nonscientists. Special thanks go to Jonathan Jones, who read the entire first draft of the translation, and to Kevin Plaxco, whose many inspired suggestions include the title of the English edition.

While I was working on this book during the night shifts, my day job took the shape of a David Phillips Research Fellowship funded by the Biotechnology and Biological Sciences Research Council (BBSRC). In a five-year project, I am now studying the folding of "newborn" proteins, and the first model protein I am looking at, the cold shock protein CspB, also appears prominently in this book. Extreme conditions seem to stay with me whatever I am doing. Maybe that's because I am an eccentric after all.

Postscript: It has been a special pleasure for me to see the arrangements for the English translation settled swiftly before the original edition was even on the market. Thus, only minor additions were needed to keep the book up to date, and a few formal changes were made to streamline the overall structure of the book. From what readers of the German version tell me and from my own impression from rereading the whole kaboodle for the purpose of translation, it should be fun to read—it certainly was fun to write and translate. Enjoy!

MICHAEL GROSS
Oxford, June 1997

1

Introduction
Life and Its Limits

Life is a paradoxical phenomenon. It is enormously varied—with creatures ranging from one thousandth of a millimeter to dozens of meters in length, having lifespans from hours to thousands of years, and with some species that spread over the whole of the globe, while others stay in their tiny ecological niche. And still, life is extremely uniform. All cells are built according to the same principles from the same molecular building blocks, no matter whether they are free-living microbes or a tiny part of a huge organism, and regardless of their position in the big family tree of life.

Life is almost everywhere, as long as we stay on the Earth's surface. However, if we consider the whole planet from its center to the outer reaches of its atmosphere, the biosphere will appear as a wafer-thin layer between the boiling lava and the freezing stratosphere. And if we were to "ask a learned astronomer" where to find life, the answer would be: almost nowhere.

Seeing that life is ubiquitous on the Earth's surface but extremely rare in space, we conclude that the conditions under which life can persist fall in a very limited range. Presumably this is even truer for the conditions under which life can arise. What makes our planet such a hospitable place is the fact that the environmental conditions are rather unspectacular compared with the rest of the universe, and amazingly constant over time. In fact, the observation that the average temperature of the Earth's biosphere has barely changed during the three and a half billion years since the origin of life, although the energy intake from the Sun was 30 percent less than it is now, led James Lovelock to formulate the much-disputed "Gaia hypothesis." His theory, named after the ancient Greek Earth goddess, states that the entire planet, including the biosphere as well as the inorganic constituents of the geo- and atmosphere, is a self-regulating cybernetic system, like a hyperorganism. We will have a closer look at Gaia in Chapter 6.

Despite the surprisingly moderate and constant climate, conditions in many areas of our planet are extreme by biological standards. In many places, microorganisms can only survive with the help of adaptive mechanisms against high or low temperatures, pressure, or chemical stress that evolved over millions of years. The number of different species in such extreme biotopes is typically much lower than in moderate habitats, and multicellular organisms are often absent. Exploring how the adaptation of microorganisms works and where it encounters its limits, we will also learn something about life, its fundamental principles, its origin billions of years ago, its diversity, and its limits.

And it is not just naturalists who are interested in microbes resisting extreme conditions. Engineers will appreciate that extremophiles may operate under high temperature and/or high pressure conditions that are quite common in chemical factories, but would be fatal for normal bacteria. Enzymes from hyperthermophilic bacteria, for instance, can be used in biotechnological processes at 100°C or beyond. Technological applications of these possibilities are only just beginning to be explored. In Chapter 3 we will discuss some of the most promising current approaches.

However, before we deal with the extremists, their habitats and limits, and their potential usefulness and danger for humans, we should recapitulate the basic facts of ordinary life: the things we need for a living, what we mean by normal after all, and which environmental conditions may limit the spread of life.

Things One Needs for a Living

A living organism is something extremely improbable, and not only because you hardly find any in the universe at large. Mainly it is improbable because it represents a highly complicated, ordered structure, while the fundamental laws of thermodynamics (the science of energy conversions) tell us that disorder of the universe (or any other closed system) must increase with time. The secret of living beings—why they can exist at least for a given lifespan despite the universal trend toward disorder—lies in their energy consumption. Living beings use up energy, the production of which creates an amount of disorder elsewhere that outweighs the order created by the organism. This is why life is so dependent on energy flow. Even a very energy-efficient organism that does not move and only replicates once in one hundred years will permanently need energy to defend its ordered state against the natural tendency for disorder.[1]

Life's energy is normally (we will discuss the exceptions) obtained directly or indirectly from the Sun. Those living beings that are able to transform the energy of sunlight into chemical energy and store it, namely the green plants and certain bacteria, are called the primary producers of the food chain. They use solar power to convert carbon dioxide from the air and water to carbohydrates and oxygen (Figure 1). Other organisms use this chemically stored energy when they eat plants and digest the carbohydrates or when they use the oxygen produced by plants to "burn" their food in energy-providing reactions. Further up the food chain, other animals eat the plant-eaters, and so on. Even bacteria degrading crude oil are using the chemical energy that plants stored millions of years ago. Similarly, if we burn fossil fuels, such as coal, oil, or natural gas, we reverse the photosynthesis reaction that plants carried out long ago. We burn the carbon that the plants built into organic molecules to form carbon dioxide and use up the oxygen produced by plants.

Next to energy provision, the availability of liquid water is the most important requirement for life. As we will see in Chapter 2, the limits of life are often defined by the boiling and freezing points of water, which under some circumstances (salty water, pressure) may be significantly different from the standard temperatures of 100°C and 0°C, respectively. The inability of other planets in our solar system to support life can largely be ascribed to the fact that they lack oceans.

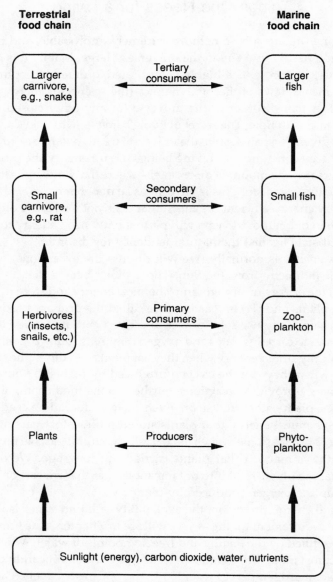

Figure 1. Two examples of simple food chains. Such chains always start from an energy source (usually the Sun), which is employed by the primary producers for the generation of biomass.

It must be said that water is not just another liquid. Most probably there is no other small molecule that combines so many special properties as the simple molecule H_2O. First of all, if the molecule behaved according to the properties of its constituent atoms according to their positions in the periodic table of the elements, the substance would not be liquid at all—it would evaporate at $-70°C$ and freeze at $-100°C$. It should be more volatile than the heavier analogue, hydrogen sulfide (H_2S), which is well known to be a gas at ambient temperature. Behind this apparently paradoxical behavior hides a phenomenon that is immensely important for the molecular architecture of all living beings—the hydrogen bond. This weak interaction links the molecules in liquid water to form an extensive network (Figure 2), implying that the chemical formula H_2O is an oversimplification. To appreciate the impact of these weak interactions, just re-

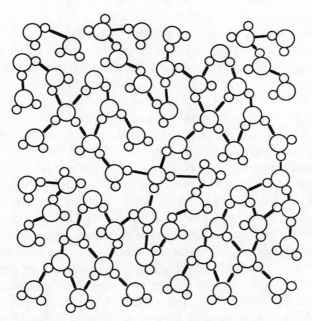

Figure 2. Schematic representation of the hydrogen bonding network in liquid water. The interaction relies on the electrostatic attraction between a positively polarized hydrogen atom (small circles) and the negatively polarized oxygen atom (large circles) of a different molecule.

member that they increase the boiling point of water by 170 degrees above what it would be if water were just a collection of H_2O molecules.

It is not only that this liquid should lawfully be a gas; it also exhibits further peculiarities in its behavior. When water cools, its density increases—as with most other liquids. At lower temperatures, molecules move more slowly and therefore need less space. However, if water is cooled to 4°C, the effect is reversed: Further cooling will decrease the density, and freezing will promote this. Both effects have important biological consequences. If, for instance, a lake cools in winter (losing most of its heat through the surface), cold water from the surface will sink and ensure thorough mixing and uniform cooling as long as the temperature is above 4°C. Below this threshold, however, cold water is lighter than the rest and will stay on top. Similarly, when the lake starts freezing over, the ice is lighter than the water and floats. Thus, the bottom of the lake can preserve the temperature of 4°C during weeks of frost, isolated by the top layers of ice and cold water. The anomalous behavior of water thus facilitates the survival of life in the lake even in severe winter conditions.

Moreover, the expansion of water upon freezing makes porous stones crack when they are soaked and then frozen. This effect, which we tend to regard as highly undesirable when it affects roads and bridges, is immensely useful in nature. It facilitates the weathering of rocks and thus speeds up the formation of soils suitable for plant growth. The scarfaced looks of the Moon and our neighbor planet Mars, which retain marks of meteorite impacts suffered billions of years ago, remind us that the relatively rapid turnover of the surface is a very special feature of our planet.

All living organisms consist mainly of water and need water to survive. The molecular building blocks of their cells need water to assume their correct functional architecture. Life is not only endangered when water freezes or evaporates. Chemicals such as salt ions can compete with cells or biomolecules for the proximity of water molecules and thus exert a chemical stress that only a very few microbes can cope with (Chapter 2).

A further special and life-enhancing property of our planet is the presence of a variety of more than 80 chemical elements. Among the 15 most common elements of the Earth's crust (Figure 3) we find a wide range of chemical properties opening a plethora of possibilities. Except for the rare noble gases, all eight groups of the periodic table are represented, and in addition we find three transition group elements (titanium, manganese, iron). In contrast, the universe at large is dominated by the two lightest

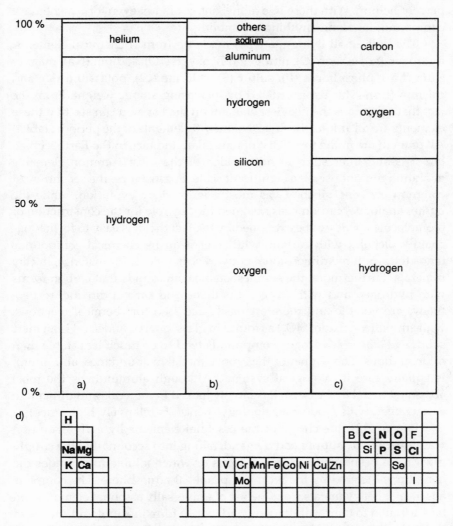

Figure 3. Relative abundance of the chemical elements (a) in the universe (atom percent), (b) in the Earth's crust (weight percent), and (c) in the human body. Panel (d) shows the elements universally required for cellular life on Earth and their positions in the periodic table; boldface indicates the 11 elements universally essential for life as it has evolved on Earth.

kinds of atoms. Roughly 90 percent of all atoms are hydrogen, another 9 percent helium. With these two alone, one could not even form a chemical compound, let alone anything resembling life.

The bulk of all living matter is made up from 11 chemical elements: hydrogen (H), carbon (C), nitrogen (N), oxygen (O), sodium (Na), magnesium (Mg), phosphorus (P), sulfur (S), chlorine (Cl), potassium (K), and calcium (Ca). (This list is ordered by increasing atomic weight. From the fact that calcium is the heaviest element on the list, you can see that these elements are all among the first 20 of the 118 elements of the periodic table.) All of them are in the top 20 list of natural abundance in the Earth's crust, but it is still interesting to note which of the most common elements evolution did not use on a significant scale. For instance, the second most common element, silicon, was mostly ignored by evolution, while the lighter analogue, carbon, was assigned the key role for the construction of biomolecules. This can be explained by the fact that it is easy to form long-chain molecules with carbon, while silicon in its chemical compounds tends to form three-dimensional networks, as it does, for instance, in clay minerals. Furthermore, the simplest stable compounds that carbon forms with hydrogen and with oxygen (methane and carbon dioxide, respectively) are volatile and therefore easily accessed for chemical reactions, while the silicon oxide (SiO_2) as found in glass, quartz, and sand is an inert solid, and silicon–hydrogen compounds tend to be much less stable than hydrocarbons. The winner of the bronze medal in abundance, aluminum, is equally rare in biological systems. Although aluminum is the most abundant metal in the crust, it is largely locked in minerals and difficult to extract even with modern technology (hence its relatively high price).

Presumably, the choice of the essential elements by early evolution was governed by supply and demand, taking into account that the supply must have been available in a gaseous or water-soluble form under the conditions dictated by the reducing primeval atmosphere. The complete absence of heavy metals from the list of universally essential elements can be attributed to their well-known tendency to form insoluble sulfides with the hydrogen sulfide abundantly supplied by volcanoes. One could, however, imagine a different outcome, within limits. If potassium had been absent when life originated, rubidium could have done the job as well. Evolution could have used manganese instead of magnesium, or strontium rather than calcium, and could still have built cells that would have looked like the ones we know and would have populated the globe quite as efficiently. The essential elements, therefore, are not essential for life as such, but have become essential for life on our planet during its evolution.

Still, it is difficult to conceive of a diverse biosphere without carbon or nitrogen. Although researchers have speculated that the first genetic information was propagated in clay minerals before cellular life arose (Chapter 5), such forms of "inorganic life" would have very limited potential to spread and adapt. If they ever existed, the "genetic takeover" by the more versatile organic molecules and cellular life has clearly marked their limitations. Thus, when exploring for life on Mars or further afield, scientists tend to look out for signs of carbon-based metabolism (Chapter 6).

In addition to the 11 universally essential elements, evolution has recruited many more for services in certain families of organisms. The heavy metals iron and zinc are essential for animals, for instance. Some bacteria require copper-containing enzymes, others incorporate selenium into their proteins, and there are many more examples of elements with special biological tasks. For the organisms concerned, these elements are of course as essential as the ones quoted earlier. However, these remain special cases that cannot be generalized to life on Earth.

Finally, life needs living space, preferably with a reasonably stable climate. The diversity of the biosphere owes a lot to the diversity of the biotopes and the accompanying opportunities for life forms to invade new habitats and adapt to the new requirements until they differ from those that stayed behind. The slow geological processes that continuously turn over the Earth's crust, such as continental drift or the formation of mountain ranges, have an important role to play in this, as they separate areas that were adjacent, create new biotopes, and change environmental conditions slowly enough to give organisms time to adapt. Once again, we are reminded of the interrelationships of bio- and geosphere which are the focus of the Gaia theory (Chapter 5).

With the diversity of biotopes, we have almost arrived at the topic of this book, but before we set out to explore the extreme biotopes of our planet, we have to face a question.

What Do We Mean by "Normal" after All?

The survival of an organism, a population, or a species may depend on numerous factors of widely different nature. First of all, there are biotic as well as abiotic factors. The former, such as predators, parasites, and competitors, are the subject of the research field of ecology, which I will not attempt to present in any detail. The latter arise from the nonliving part of the environment, from physical and chemical conditions that are directly related to the forms of stress that this book is about. In turn, they can be classified into physical and chemical factors.

Let us start with the simple things. Some physical parameters are practically constant on the whole surface of our planet. The atmospheric pressure at sea level is approximately one atmosphere (1.013 bar), wherever you are. The variations in atmospheric pressure that we experience as the changes in weather only amount to a few percent of this constant. Earth's gravitational force is the same everywhere. Each falling body is accelerated by 9.81 meters per second every second of its fall (if we ignore friction).

The chemical composition of the atmosphere and the oceans is rather uniform as well. The freezing air over the poles as well as the steaming air of the jungle contain 21 percent (by volume) oxygen, 78 percent nitrogen, with carbon dioxide and noble gases sharing the remaining 1 percent— provided one removes the variable content of water vapor. Seawater always contains about 3.0 percent salt, the rivers and lakes much less (except for lakes with no outflow, like the Dead Sea).

So much for the global constants. The most striking variability is observed in temperatures. They range from −80°C in the antarctic winter to 250°C in the plumes of deep sea hydrothermal vents. As most organisms do not enjoy the comfort of an inbuilt central heating system the way we mammals do (to say nothing about insulation standards), the "body temperature" is, for most cells, exclusively governed by the environmental temperature.

Which temperature you regard as normal depends strongly on whether you were born and bred an Inuit or a Californian, a polar bear or a gut bacterium. The latter, if they dwell in human intestines, would regard 37°C as the most normal temperature in the world. This meets the approval of all medically or zoologically oriented biochemists and microbiologists, who define physiological conditions as a lukewarm, slightly salted brine. Researchers who favor practical considerations would define normality by the standards of the common or preferred room temperature, for instance 20 or 25°C depending on the altitude of their habitat.

However, there is no such thing as a norm or a physiological standard temperature for all living beings. At best, one could define a range of moderate temperatures as opposed to extreme temperatures. Thus, a range from 20 to 40°C would be a normality that could be accepted in a global opinion poll. Organisms that prefer warmer temperatures than these will henceforth be called "thermophilic," and those that prefer cooler temperatures "psychrophilic."[2] But we will realize that for many living beings much higher or lower temperatures are perfectly normal living conditions.

In the oceans, hydrostatic pressure (which increases linearly with the depth under the surface) joins temperature as another physical variable from which the cells cannot isolate themselves and to which they must therefore adapt. As we will see, our human view that one atmosphere should be called the normal pressure is factually incorrect for more than half of the biosphere. Three-quarters of the total volume of the oceans, corresponding to 62 percent of the biosphere, is subjected to hydrostatic pressures more than a hundredfold higher than our beloved atmospheric pressure.

With respect to the chemical variables, the average, most common values are normally regarded as standards. Neutral pH (neither acidic nor alkaline), moderate salt concentrations, availability of oxygen and nutrients— these are typical conditions found in the more hospitable parts of the Earth's surface. One could call those conditions "normal" or "physiological," which allow the archetypical "Joneses" of microbiology, the gut bacteria called *Escherichia coli* (or *E. coli* for short), to thrive. But there are quite a lot of less hospitable places with less "physiological" conditions as well. These exceptions to the rules will teach us something about life's capacity for adaptation and the limits of life.

The Limits of Life on Earth

Even though life on Earth has, in the course of evolution over billions of years, adapted to various extreme conditions in amazing ways, as we will discover later, the laws of physics sometimes put a halt to adaptation and define an absolute limit beyond which no life can exist. With respect to heating and cooling, the temperature as such is much less limiting than the availability of liquid water. In environments where salt content and/or hydrostatic pressure increase the boiling point and decrease the freezing point of water, extremophilic microorganisms can thrive at 110°C or at −5°C, as long as the water remains liquid. Of course, the range cannot be expanded indefinitely. Research into the stability of biomolecules suggests that an upper temperature limit exists at around 120°C, beyond which the molecules of the cell would decompose more rapidly than the cell could replace them.

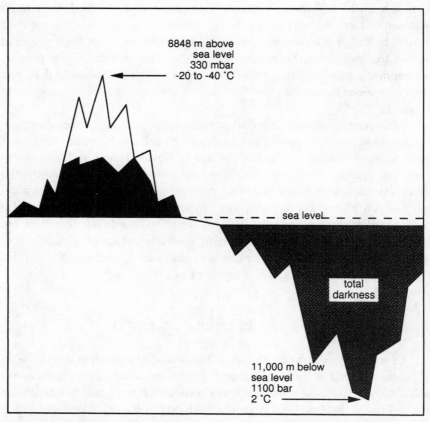

Figure 4. Schematic representation of the greatest heights and depths on Earth, with the corresponding values of pressure and temperature.

As for the hydrostatic pressure in the deep sea, an upper limit of pressure tolerance of cells certainly exists, but the oceans are not deep enough to reach it. Even at the highest pressures occurring in the deepest trenches of the Pacific (1,100 bar at 11 kilometers below surface), bacteria can thrive. Jules Verne, in *20,000 Leagues under the Sea*, predicted correctly that one would find life at these pressures. However, he overestimated the depth of the oceans and assumed a dead zone beyond 12 kilometers. (A schematic representation of the greatest heights and depths of the earth with corresponding temperatures and pressures is shown in Figure 4.) The fundamental limit set by physics may lie somewhere between twice and three times the highest pressure found in the oceans. Of course no one knows the limiting pressure to which organisms could have adapted if there had been deeper oceans to populate.

Chemical stress, such as high salt concentration, or acidity, does not set firm limits to life. As we will see later, in the case of acidity this is partly due to the fact that this stress factor can be excluded from the interior of the cell, requiring adaptation only for those parts that deal with the outside world.

In summary, we should acknowledge that our home planet is quite a nice place to live. The relatively mild and constant conditions have encouraged the evolution of a biosphere of enormous diversity. We will get an impression of how diverse life on Earth is when we visit its fringes, those extreme biotopes that evolution may have had to conquer during millions of years of adaptation to extreme conditions.[3]

Endnotes

1. The eminent physicist Erwin Schrödinger (1887–1961) addressed this issue— the thermodynamics of the living cell or how to create order from disorder—in his famous 1943 lecture series, "What Is Life?" However, it was the issue of "order from order" which had direct influence on the founding of the science of molecular biology.
2. From ancient Greek *thermós, psychros*, and *phílos*, for warm, cold, and friend, respectively.
3. However, some researchers believe that the history of life began with extreme conditions, specifically with hyperthermophiles in hot submarine biotopes (Chapter 5), suggesting that we mesophilic organisms lost the heat resistance when we settled for a cooler lifestyle.

2

Extreme Environments and Their Inhabitants

You may have read sensationalist headlines claiming "life in boiling water," "stone-eating bacteria," "bugs at the bottom of the sea," or something similar, and you may have wondered, are these actually true? Well, although few of these things were known or even suspected a couple of decades ago, scientists now have compelling evidence for all of these claims and for many others that are equally amazing. Inhospitable places on our planet, from steaming hot springs to the deep freeze of the polar regions, tend to be populated by specially adapted microbes that seem to enjoy the specific extreme conditions and that are therefore called "extremophiles." In this chapter we will explore a variety of sites where environmental conditions range from the uncomfortable to the downright nasty and will meet the extremophilic microbes that live there.

profile _____

Thomas Brock and the Discovery of the Hyperthermophiles

Thomas D. Brock was born in Cleveland, Ohio, in 1926 and admits to being self-taught in almost all of his many occupations; he had already been on an odyssey through various branches of biology before he visited Yellowstone National Park for the first time in 1964. At the time, he was particularly interested in the ecology of microorganisms, and hence he was surprised and fascinated to find rich microbial life in the outflow channels of the hot springs, in the form of colorful mats or pale-pink gelatinous masses. The next summer he and his wife, Louise Brock, returned with scientific equipment for a two-week "working vacation." They first studied the algae that lived in the outflow channels at greater than 60°C. But then they found evidence that bacteria flourished even in springwaters as hot as 82°C. They isolated and described the first extremely thermophilic (hyperthermophilic) bacteria. Among the very first isolates was one Brock baptized *Thermus aquaticus*. More than two decades later, the DNA-synthesizing enzyme of this bacterium (Taq-polymerase) became a best-selling biochemical, as we will see in Chapter 3.

Why did nobody else investigate these bacteria before Brock, as Yellowstone is well known and open to everyone? There is a simple reason: In order to isolate a certain species of microorganism, microbiologists carry out so-called enrichment cultures. First, they take a small sample of soil or water home to their laboratory, which they then cultivate on sterile growth media under conditions that they think favorable for the organism they want to find. The outcome of this approach is of course heavily influenced by fundamental assumptions or prejudices. If researchers believe (as they actually did before Brock) that 55°C is an extremely high temperature for any microbe, and if they therefore grow their enrichment cultures for the isolation of thermophilic bacteria, say, at 40, 50, and 60°C, they will never find a bacterium that has its growth optimum at 95°C. Thomas Brock, however, had always been in favor of field studies and observation of microbes in their natural environment. So he took samples of the hot springs, cultivated them in media that mimicked the composition of the spring water, and sometimes even used the very same spring as a thermostatic bath for the incubation. And with this straightforward approach, he opened up a whole new world of microbiology: To put it in a nutshell, Brock found life in boiling water.

Brock became famous, both as the founder of the research field of hyperthermophiles and as the author of the classical textbook *Biology of Microorganisms*. In a recent autobiographical essay, he specifies the exact moment when he realized that he had made it. It was in May 1970, at a cocktail party of the American Society for Microbiology. When he mentioned that he worked at the University of Indiana, he was asked: "Are you with Brock?" Truthfully, he replied: "No, I *am* Brock." Meanwhile, he has a bacterial species to his name (*Thermoanaerobacter brockii*) and has become a role model for generations of microbiologists who started looking for life in apparently inhospitable places and exclude the word "impossible" from their vocabulary.

Some Like It Hot: Life around Geysers and Volcanoes

Terrestrial Hot Springs

The discoveries of extremely heat-loving (hyperthermophilic[1]) bacteria made by Thomas Brock and his co-workers in the hot springs of Yellowstone National Park during the 1960s (see Profile) encouraged other microbiologists to hunt for hyperthermophilic microbes in the hot springs and boiling mud holes of other volcanic regions in other continents and on the sea floor. Suitable places are found, for instance, around the Pacific, in Iceland, and in Italy. The hunters and gatherers of extremophilic microbes, such as the German microbiologist Karl Otto Stetter, travel regularly to the geysers of Iceland, the solfatara fields in Italy, or the volcanoes of New Zealand. In all these places, water is heated by volcanic activity, often up to the boiling point.

Most (but not all) hot springs are found near active volcanoes. They work by the same principle as an electric shower, namely by heating the water continuously while it flows through (Figure 1a). Groundwater sifts through cracks in the ground down to the volcanically heated rocks, from where it rises to the surface through different channels.

Geysers, in contrast, are more like a reservoir boiler, which heats a huge amount of water discontinuously. The reservoir has only a single channel for entry and exit, and the water, which is heated from below, cannot circulate freely (Figure 1b). Depending on the height of the water column and the hydrostatic pressure it generates, the boiling point of water is increased at the bottom of the geyser. Thus, the overheated water can reach temperatures of, say, 120°C. When it eventually does start to boil (Figure 1c), the steam presses part of the water out of the "bottleneck," thus reducing the liquid column and the hydrostatic pressure. This enables a bigger fraction of the overheated liquid to evaporate rapidly. The positive feedback in this process leads to a rapid discharge of the whole content of the reservoir, observed as a spectacular fountain above ground (Figure 1d). After that, nothing seems to happen for some time (one to several hours) while the reservoir is refilled and the fresh water heated to the boiling point.

Solfatara occur in areas of fading volcanic activity. They are gas excretions of the volcanic soil, which contain (as the name suggests) sulfuric gases, mainly the malodorous hydrogen sulfide (H_2S). Investigations of the hot, moist soils of solfatara fields in various places have shown that these always contain two distinct layers. The upper 15 to 30 centimeters is ocher-colored due to the presence of iron oxides and is characterized by the presence of oxygen as well as high acidity. These fields often contain boiling hot mud or water pits, with the same chemical conditions. Although they do not look very hospitable, these volcanic soup kettles often

---→

Figure 1. Hot springs and geysers: Rough schematic description of their basic principles. (a) Hot springs occur in places where groundwaters continuously pass through volcanically heated layers of rocks and then rise to the surface through separate channels. Geysers, in contrast, work discontinuously as their "boiler" is filled and emptied through the same channel. (b) The hot area is filled with fresh water, which, however, cannot escape immediately. (c) This water is heated to boiling. (d) When the overheated liquid at the bottom of the "boiler" begins to boil, the evaporation spreads explosively and discharges the whole water content in one go.

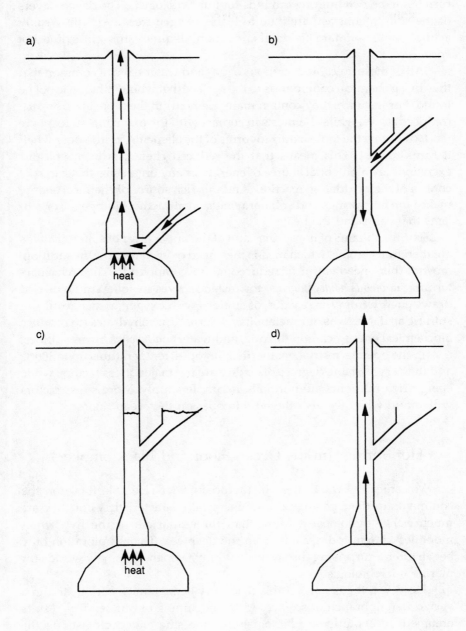

turn out to be rich hunting grounds for microbiologists. The deeper layers of the soil, in contrast, are blue to black, oxygen-free, and only weakly acidic. Some solfatara fields also contain alkaline, strongly saline hot springs.

All of these volcanic excretions of gas and water have in common that they bring chemical compounds to the surface that would otherwise not be found. For instance, they contain many elements in the more electron-rich (reduced) form, while chemicals in contact with the oxidizing atmosphere tend to assume the more oxidized forms of the elements, as iron does when it turns to rust. This means that the reduced chemicals from volcanic excretions are a potential source of energy for any organisms that can take control of their oxidation reaction. And this rich source of chemical energy makes the hot springs and solfatara fields an attractive biotope for organisms that can stand the heat.

The inhabitants of hot springs and of the upper layer of solfatara fields are therefore aerobic, which means they need oxygen to grow. In addition, most of them need sulfur or reduced sulfur compounds. The archaebacterium *Sulfolobus acidocaldarius*, for instance, lives in solfatara fields and draws energy from the reaction of sulfides with oxygen, which results in sulfuric acid. Much as our metabolism "burns" carbohydrates (to produce biochemical energy, carbon dioxide, and water), *Sulfolobus* burns sulfides.

In the open terrestrial geothermal biotopes, temperatures up to 100°C and the very variable degrees of acidity are the major stress factors, while energy-rich nutrients are normally in ample supply. More stress factors will join in when we visit the hot biotopes on the ocean floors.

Hot Springs on the Ocean Floor and Black Smokers

Volcanically heated areas of the ocean floor resemble those on the continents in their geological and chemical characteristics, but they are much richer biologically. More than three-quarters of the hyperthermophiles known today come from the deep sea. Their exploration only began after a surprising discovery that in 1977 opened up new worlds for marine microbiologists.

However, the history of this discovery begins much earlier, and in a very different branch of science. At the beginning of our century, results obtained by evolutionary biologists on one side and geologists on the other seemed to contradict each other. The evolutionary relationship between plants and animals living on separate continents far apart from each other required primeval land-bridges or now-vanished intervening continents as an explanation, while geological investigations demonstrated that continents are stable and cannot sink without a trace—whatever some people may claim about the legendary Atlantis.

A man named Alfred Wegener[2] dared to think the unthinkable. If continents must have been linked in the past, and if the linkers cannot have disappeared, then the continents must have moved. In his first publication on "The Origin of the Continents" (1912), he put forward the continental drift hypothesis and suggested that one should "regard the mid-atlantic ridge as a zone where in the course of the continuing growth of the atlantic ocean its floor is continuously torn open and gives way to fresh, relatively fluid and hot sima[3] from the depths." As we know now, this description was spot on, but more than half a century would pass before this revolutionary idea was recognized and confirmed through direct measurement of the continental drift (Figure 2), which amounts to several centimeters per year.

In the 1950s, Wegener's hypothesis was put onto more solid (if still hypothetical) geophysical foundations, the theory of plate tectonics. This

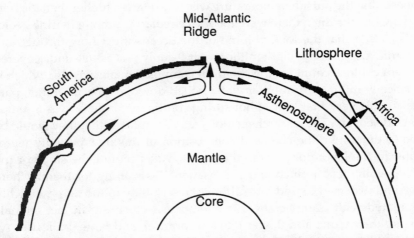

Figure 2. Schematic cross-section of the Earth showing the slow movements which cause Africa and South America to drift apart.

states that the Earth's crust (the outermost layer of the earth, reaching 25 to 40 kilometers deep) consists of roughly 10 plates that move relative to each other. Both plate tectonics and continental drift are now considered as proven facts. We know the places where plates slide on top of others, drift apart, or brush alongside each other. Most interesting for geologists are the cleavage zones where plates move away from each other and new crust is formed through the extrusion and cooling of fresh lava.

In 1975, researchers were for the first time able to detect fresh lava at the mid-Atlantic cleavage zone. When they tried to discover further indications of the movement of the continental plates by suspending a camera above the sea floor of the Pacific at a depth of 2,500 meters in a geologically active region north of the Galapagos Islands, they discovered surprising turbidities in the seawater, as well as unusual white "sediments."

In order to inspect these mysterious apparitions more closely, the geologists John B. Corliss (Oregon State University) and John M. Edmond (Massachusetts Institute of Technology) boarded the research submarine *Alvin* in the spring of 1977. The vessel, which can take a pilot and two researchers to a maximum depth of 4,000 meters below sea level, is operated by the Woods Hole Oceanographic Institution, which was founded in 1930 in the seaside village of Woods Hole, Massachusetts, and is now one of the biggest and most renowned centers of ocean research in the world. When they reached the slope of the midoceanic ridge, the geologists first noticed that the outside temperature was five degrees higher than the normal 2°C. At the time, there were theories circulating among marine geologists stating that the total volume of the oceans must flow through hot volcanic rocks once in eight million years—only this as-yet-undiscovered process could account for the chemical composition of seawater, which is drastically different from river water boiled down in an evaporation pan. Hence, this very first hint of hot springs on the ocean floor was a sensational discovery. The researchers took water samples so that they would be able to determine the chemical composition of this unexpectedly warm water in the laboratory. Then, they made *Alvin* mount the slope of the ridge. At the top, a much bigger sensation was waiting for them. Where they had expected to find a basalt rock desert, they found an oasis richly populated with clams, crabs, sea anemones (which are in fact animals despite their name and flower-like appearance), and large, pink fish. As Edmond later recalled in *Scientific American*, they spent the remaining five hours of their dive in frantic excitement. They measured temperatures, conductivity, pH, and oxygen content of the seawater, took photographs, and collected specimens of all the animal species. As it turned out, the researchers had discovered an entire field of warm springs. On an area about 100 meters in diameter, relatively warm water of up to 17°C was sifting through every little crack of the sea floor.

Holger Jannasch, a German marine biologist who has been working as a senior scientist at the Woods Hole Oceanographic Institution from 1963 to this day, was one of the first to hear the news. He recalls that he "got a call from the chief scientist, who said he had discovered big clams and tube worms, and I simply didn't believe it. He was a geologist, after all." Later on, Jannasch had ample opportunity to witness the marvels of the deep sea oases with his own biologist eyes and study the organisms in detail. And in spring 1979, another dive of the *Alvin* brought new sensations: hot springs that eject thermal fluid at 350°C into the ocean.

These hydrothermal vents, affectionately known as "black smokers" to their friends, are hot springs in geologically active regions of the ocean floor (Figure 3). There are a few distinctions with respect to their terrestrial cousins. First, the hydrostatic pressure at depths of around 3,000 meters is 300 atmospheres, which is enough to increase the boiling point of water beyond 400°C. Hence, "hot water" can be really hot on the ocean floor, where the normal temperature is 2°C. Second, the hot water is ejected not into the air, but into cold seawater. The rapid mixing of the overheated volcanic fluid with cold water leads to the instantaneous precipitation of several substances (mostly heavy metal sulfides) in the immediate vicinity of the outflow site. These precipitates form the characteristic chimneys, which can reach heights of up to five meters. Further precipitation reactions lead to the typical black "smoke" that has given the vents their nickname. Around 30 regions with such vents have been discovered. Even the death of a black smoker and its rebirth at a nearby site was observed "live" as it happened.

The chemical composition of the water ejected by a black smoker matches exactly the predictions of marine geologists. Due to reactions of the seawater with hot basalt deep under the sea floor, the thermal fluid is poorer in magnesium and sulfate, but richer in iron, manganese, and sulfide than ordinary seawater. The discovery of black smokers instantaneously put the material balance of the oceans right, the apparent imbalance of which had worried geologists for decades. To appreciate the importance of the newly discovered exchange process, one has to realize

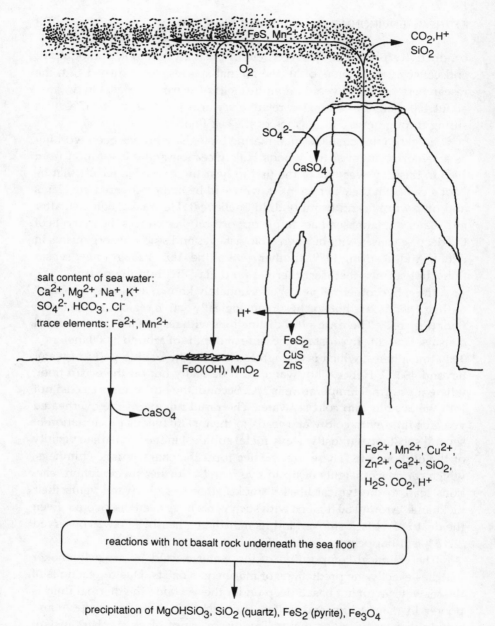

Figure 3. Schematic cross section of a black smoker, with the reaction paths of the most important minerals involved in the formation of the chimney and the precipitation producing the "smoke" plume. The chemical elements involved are: C, carbon; Ca, calcium; Cl, chlorine; Cu, copper; Fe, iron; H, hydrogen; K, potassium; Mg, magnesium; Mn, manganese; Na, sodium; O, oxygen; S, sulfur; Si, silicon; Zn, zinc.

that the total volume of the oceans indeed passes through the hot basalt rock every eight million years. For geologists, this timespan is a mere moment, and it means that every drop of water in the oceans has gone through deep sea vents hundreds of times since our planet's surface consolidated.[4]

Surrounding the smoking vents, researchers found wonderful worlds of unknown creatures resembling the biotopes discovered at warm sea-floor springs two years earlier. More than 370 new species have been discovered in deep sea thermal biotopes, and more than 90 percent of these are exclusive to these biotopes. Population densities in these pitch dark oases exceed those of sunny, nutrient-rich coastal waters. Mollusks, tube worms, sea anemones, snails, and crabs thriving in the lukewarm waters surrounding the vents and sources would not stand a chance of surviving in the dark and cold deep sea if it was not for the presence of the thermal waters.

The question of how species can spread to populate several of these small biotopes, which lie far apart and are unreliable on a longer time scale, is a tricky one. In 1996, Verena Tunicliffe and Mary Fowler of the University of Victoria (Canada) suggested that the animals had learned their lessons in plate tectonics and somehow brought their larvae to migrate preferably alongside the midoceanic ridges. The pattern of species spread over the globe is most easily explained on the basis of the assumption that certain pathways leading through the inhospitable deep sea were taken with lesser probability than those that may have been longer, but followed plate boundaries and were more likely to meet new oases at regular intervals. If confirmed, this finding would suggest the interesting possibility of using the evolutionary relationships between the inhabitants of geothermal habitats to detect ancient plate boundaries, the geological traces of which may have vanished already.

However, the cozy temperatures alone cannot explain the wealth of the fauna in the surroundings of hydrothermal vents. After all, at 3,000 meters below sea level, photosynthesis is out of the question, and the nutrient supply raining down from the upper layers cannot be called generous. Hence, there would not be enough energy to feed the deep sea biotopes if there was not another source.

Investigations into the ecology of the hydrothermal vent communities have shown that all of the multicellular organisms found in these biotopes depend on thermophilic bacteria, which are the only organisms that know how to draw energy from the reduced sulfur compounds contained in the thermal waters. This special kind of metabolism was discovered by Russian microbiologists more than a century ago in the bacterium *Beggiatoa*, but was not given very much attention at the time. In the case of the tube worm, *Riftia pachyptila*, the dependence has developed into an intracellular symbiosis. The worms, which have neither mouth nor gut, live exclusively off the nutrients that the chemosynthetic bacteria produce inside them (see Figure 6 in Chapter 3). In compensation, they provide the bacteria with the raw materials, mainly hydrogen sulfide, oxygen, and carbon dioxide, in highly enriched concentrations. I will explain this symbiosis in more detail at the end of the next chapter.

Researchers think that the sparse nutrients raining down from the photosynthesis-dependent food chain in the top layers of the oceans can only contribute a minute fraction of the energy requirements of the abundant black smoker communities. Instead, there is an independent food chain, which starts from the sulfur-oxidizing bacteria as the primary producers and leads to the crabs, mollusks, and fishes as consumers (Figure 4). Being hyperthermophilic definitely is an advantage for the primary producers of this newly discovered food chain—the hotter the "soup" they drink, the higher the concentration of the poorly soluble, energy-rich sulfides in it. Therefore, it is easy to understand how evolution favored hyperthermophiles and why the record-breaking extremists are found in black smoker communities.

There have also been false claims of high temperature records. One group claimed in 1983 the discovery of bacteria that live at 250°C within the thermal fluid of a black smoker. However, other groups were quick to demonstrate that at this temperature almost all biomolecules fall to pieces within seconds. The error was most probably due to the sampling method of the former group. They "capped" the vent for a short period with a container to take a sample of the thermal fluid. However, the rising of the hot thermal fluid also induces a convection movement, which sucks colder water from the surroundings into the container. Most probably, this was the way in which the alleged super-hyperthermophiles entered the samples.

Still, the proven and reproducible resistance of microorganisms against temperatures beyond 110°C is fascinating enough, and it occurs only in certain branches in the family tree of life. Of the more than 20 genera of hyperthermophilic species, only two, namely *Thermotoga* and *Aquifex*, belong to the domain of the ordinary bacteria (eubacteria). The first representative of the genus *Thermotoga*, the bacterium *Thermotoga maritima* (Figure 5a), was first discovered in the Mediterranean, off the shores of Italy. Nowadays, however, it can be more easily found in the laboratories of the University of Regensburg, Bavaria, where researchers

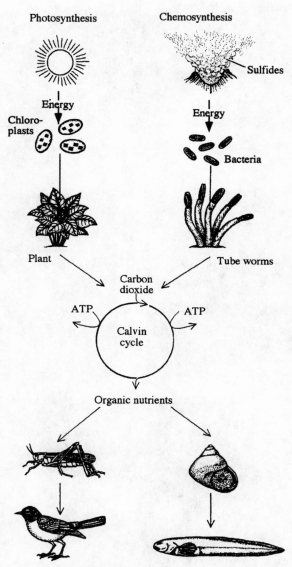

Figure 4. Comparison of the food chains based on photosynthesis and chemical synthesis. Both lead to the same final products. They mainly differ in the energy source, which is the Sun in one case, the oxidation of sulfur compounds in the other. In the chemosynthetic food chain, tube worms adopt the role of the plants in the photosynthetic food chain—like these, they use carbon dioxide to build carbohydrates, lipids, and amino acids, which then serve the nutrition of higher organisms.

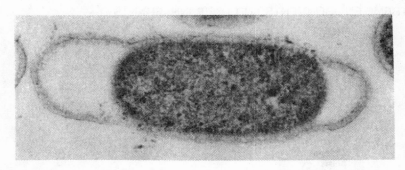

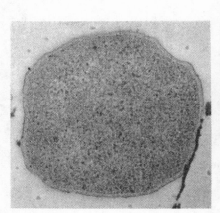

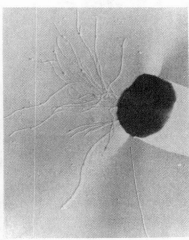

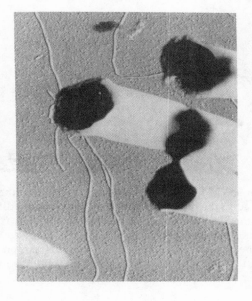

a

b

c

d

have grown quite fond of it—partly because *Thermotoga* is a rather pleasant hyperthermophile to work with, as it neither produces nor requires sulfuric gases. While the microbiologists in the group of Karl Otto Stetter, who discovered and first described the heat-loving bacteria in 1986, grow them in their special high temperature fermenters and characterize them microbiologically, biochemists in the laboratory of Rainer Jaenicke purify various enzymes from *Thermotoga* with the ultimate goal of finding the secret of molecular adaptation to high temperatures.

In contrast, the overwhelming majority of hyperthermophiles belong to the kingdom of the archaebacteria (Chapter 6), a third life form distinct from both "normal" bacteria (eubacteria) and the complex cells of higher organisms (eukarya). Among the archaebacteria, we also find the "most extreme" extremists, like the orders of *Pyrococcus*, *Pyrobaculum*, and *Methanopyrus*, whose members all grow at temperatures beyond 100°C. The current (1997) record is held by *Pyrolobus fumarii*, an archaebacterium isolated by Stetter and co-workers from the wall of a black smoker in the Atlantic, which grows between 90 and 113°C. Nobody knows, however, where the upper limit of adaptation is. Most researchers in the field believe that it lies somewhere between 115 and 150°C.

Stay Cool: Life at Subzero Temperatures

Apart from hot springs and the thin layer warmed by the sunshine, oceans are a rather cold place to live in. The standard temperature in the deep sea is 2°C, while temperatures in the polar seas can even sink below 0°C, as the freezing point is lowered by the presence of salt.

These temperatures, however, are not quite as life-threatening for cells as those near the boiling point. Molecular building blocks of the cell can withstand low temperatures quite well. Scientists can, for instance, cool bacterial cultures to −20°C in a water/glycerol mixture for storage and revive them at any time by warming them and diluting them into fresh medium. The major problem for organisms frequently exposed to temperatures around zero is how to stop the water they need to survive from freezing. Or, in the worst case, how to guide the ice formation in a way that

←——————————————————————————————

Figure 5. Electron micrographs of some extremophilic microbes. (a) The "hottest" eubacterium, *Thermotoga maritima* (0.7 by 2.2 micrometers); (b) *Methanococcus igneus* (ca. 1 micrometer in diameter); (c) *Pyrococcus furiosus* (ca. 1 micrometer in diameter); (d) *Metallosphaera prunae* (diameter 0.7 to 1.0 micrometers).

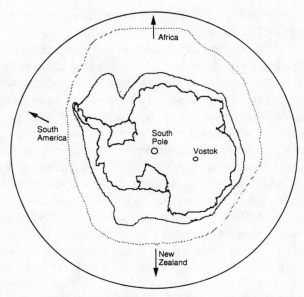

Figure 6. Map of Antarctica with the pack ice (solid line) and drift ice (dashed line) limits. The location of the Vostok research station is also indicated.

avoids damage. In Chapter 3 we will discuss various strategies to avoid "frostbite" on a cellular scale.

Much lower temperatures, however, can occur on dry land, especially in the antarctic (Figure 6). The lowest temperature ever measured on Earth was −88.3°C, recorded in 1959 at the antarctic research base Vostok. Temperatures between −40 and −50°C are quite normal on our planet's coldest continent. At the South Pole, the average temperature in summer (December) is −29°C, and in winter (August) is −61°C. In the north polar region, summer temperatures are typically around freezing, winter temperatures near −40°C.

Considering these conditions, it comes as a surprise that the antarctic harbors more than 1,000 plant species (regarding animals, see Sidelines, "Of Polar Bears and Penguins"). With only two exceptions, they belong to the cryptogams, a group of plants including ferns, moss, and lichens, which can reproduce through spores. In Chapter 3, we will discuss the special adaptations that provide spores (both plant and bacterial) with a unique ability to resist extreme conditions.

Lichens have been by far the most successful settlers on the antarctic continent. As far as simple survival of a period of bad conditions is concerned, cold cannot do them any harm. In research laboratories, lichens have survived temperatures near absolute zero ($-273.15°C$) and absolute dryness without difficulty. Furthermore, they are completely independent of their substrate—they thrive on naked rock, concrete, or glass. The secret of their success lies in a symbiotic relation between a fungus and an alga, as I will explain in more detail in the next chapter. One thing that proves limiting for the success of lichens in the antarctic, however, is the fact that temperatures stay below zero all year round. Although they can survive any cold temperature for any length of time in a passive state, they need liquid water to live an active life and to spread.

They find a set of relatively benign conditions on the sun-exposed sides of rocks in the antarctic mountain ranges. The rocks can easily reach temperatures between 0 and $10°C$, while being moist from melting snow. For lichens, whose photosynthesis already works optimally at $0°C$, these rocks are paradise. In the coastal regions, where sea birds fertilize the rocks with their guano, free-living (i.e., nonlichenized) algae are observed along with lichens, for instance, the green alga *Prasiola crispa*.

At the shores of the antarctic continent, a swimming ice shield grows to a size that would suffice to cover Europe (20 million square kilometers) every winter, four-fifths of which disappears in the following summer. Even within this ice, which is up to two meters thick, plant life can be found. When the first thin layer accumulates from small grains of ice, marine algae get trapped in cavities. Some algal species not only survive this for a winter, but can even produce biomass while stuck in the ice shield. Paradoxically, life is hardest in the upmost third of the ice layer. It is colder than the rest, so the remaining enclosures of liquid water are smaller and saltier, as the sea salt cannot take part in the formation of proper ice crystals. This is why life can only rarely be found in the surface layers of the ice shield.

In the middle layers (80 to 120 centimeters below the surface), the frost is moderate enough to allow cold-adapted algae some sort of normal life. Even small animals feeding on algae can survive here. In the lower third of the ice shield, the lack of light becomes the limiting stress factor. At the bottom of the ice shield, of course, the ecosystem of the ice meets that of the seawater, but not very much is known about the ecology of this specific border.

As in almost all the oceans, krill is thought to play a major role here. These little animals can gobble up algae at an incredible speed and serve as an important food source for various fish and whales.

In the northern polar region, in contrast, the ice is much less "alive" than in the antarctic. One reason for this is that the polar sea is surrounded by continents, which block the heat supply from warm oceanic currents. Therefore, only a minor part of the pack ice melts during the summer, and there is little chance for algae to colonize the ice.

There are further potential biotopes in the antarctic that as yet have literally remained white spots on the map. Beneath the continental ice sheet lie dozens of freshwater lakes, whose existence is only known through satellite exploration. Most of them are rather small, but one of them, Lake Vostok, may cover a similar area to Lake Ontario, with a maximal length of 200 kilometers and an average depth of 125 meters. Although the existence of the lake was discovered in 1974, its true size was only found during a recharting using the satellite ERS-1, which measures the echoes of radiowaves.

Although the temperature in the four-kilometer-thick ice sheet covering Lake Vostok may be as low as −50°C, such lakes can persist for millennia. This paradox is partly explained by the Earth's heat, which the lakes receive from below and which they can store quite efficiently, as they are well insulated by the ice layer above. In addition, the enormous pressure generated by the weight of the ice lowers the freezing point of the lake water. This effect is well known to ice skaters. The pressure of the skate melts the ice and thus generates a thin film of liquid water, which lubricates the contact between skate and ice.

As yet, none of the deep drillings carried out in the antarctic has opened access to one of these freshwater lakes. In 1996, a drilling program of the Vostok research station stopped ca. 150 meters above the surface of Lake Vostok. Researchers are excited about the possibility of finding a biotope that may have been secluded under the ice shield for something like two million years. But there are also fears that the hypothetical biotope could be irretrievably destroyed by the very process of drilling a hole that would put it back into contact with the rest of the world. Therefore, some scientists have suggested a practice drilling into one of the smaller lakes first before Lake Vostok is put at risk.

Apart from the exciting prospect of a new biotope, the lakes underneath the ice are also regarded as terrestrial models of potentially life-

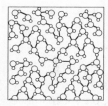 sidelines _____

Of Polar Bears and Penguins—Vertebrate Life at the Poles

Life on the antarctic shores, of course, is not limited to algae and lichens. There are penguins, albatrosses, petrels, and various species of seal (which remain relatively unaffected by the wicked deeds of humankind, as they are so far away from the crowded parts of the world). And the north polar region has the polar bear, polar fox, snow hare, lemming, reindeer, and wolverine, as well as seals and sea birds. No less than the microbes, these creatures, too, experience the polar climate as an extreme condition, to which they have adapted in the course of many generations. That their environment is far from ideal for life is already obvious from the relatively small number of species found in the polar areas.

In contrast to microorganisms, however, these animals have the option to adapt on physiological and anatomical levels rather than on a cellular or molecular one. The cells of a polar bear on the pack ice do not suffer more stress than those of a panda in a bamboo grove. The polar bear just has to invest more energy in the production of a thick insulating fur and in heating its body. Furthermore, adaptations in behavior can help animals to survive the cold. Various animals of the arctic build caves in the snow as a protection from cold winds. Breeding emperor penguins are known to stand together as close as possible in a circle and to change shifts in the colder places at the outer border.

For polar fish, however, the cold invades the body, requiring the invention of biological antifreeze agents, as we will see in Chapter 3. A positive effect of the cold is the higher solubility of gases in water, and thus also in body fluids. A family of antarctic fish, the ice fish (*Channichthyidae*), have used this effect to dispense with a costly transport system. They have very few red blood cells and no hemoglobin. At temperatures around zero, their blood dissolves enough oxygen to supply the cells, but at higher temperatures, these fish would die.

---

supporting places in the solar system. Jupiter's moon Europa, for instance, is known to be covered with a thick ice shield that looks much like the antarctic one. If life was found underneath the latter, bets would go up that there might also be life on Europa (see Chapter 6).

Living under Pressure: The Deep Sea

In 1643, Evangelista Torricelli (1608–1647), mathematician at the court of the Dukes of Tuscany—a position he took over from Galileo Galilei—invented an extremely useful measuring device. However, he could not appreciate the importance of his device immediately, as he discovered the measured quantity at the same time. His experimental setup consisted of a tube filled with mercury and sealed at one end. The open end was immersed in an open trough filled with mercury as well (Figure 7). (Don't try

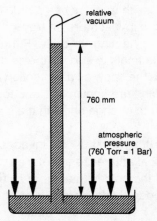

relative
vacuum

760 mm

atmospheric
pressure
(760 Torr = 1 Bar)

Figure 7. Principle of the Torricelli barometer. The atmospheric pressure is in balance with the weight of the mercury column, whose height can thus be used as a measure for the pressure.

to repeat this experiment—obviously, the Dukes of Tuscany didn't have a safety officer for their laboratories!) If one puts the tube in a vertical position, with the sealed end at the top, the mercury will sink a bit, but its meniscus will always be about 76 centimeters above the surface of the mercury in the trough. This means that the pressure exerted by the weight of the atmosphere onto the open mercury surface corresponds to that of a mercury column of 760 millimeters height, or 760 Torr, as measured in an old pressure unit derived from both Torricelli's name and his setup.

Thus, in one simple experiment, Torricelli discovered the atmospheric pressure and a (relative) vacuum (in the sealed end of the tube, above the meniscus) and invented the barometer at the same time. Despite its highly toxic vapors, mercury was the best choice for this experiment, being the heaviest substance that is liquid at room temperature. To carry out the same experiment with water, Torricelli would have needed a tube 11 meters in length, as the atmosphere exerts the same pressure as a water column of just above 10 meters in height. Officially, this standard atmospheric pressure is quoted as 1.013 bar (or, even more officially, in the extremely unwieldy unit Pascal, in short Pa, which, fortunately, relates to the bar through a simple shift of the decimal dot, as 101,300 Pa). The unit "one atmosphere" is now abolished—as are "pounds per square inch" and Torr—but can, for rough calculations, be used as an equivalent to bar, as the error produced by this replacement is only ca. 1 percent.

Considering that the atmosphere exerts the same pressure as a water column of roughly 10 meters in height, you can deduce what happens when you dive down into the ocean. For every 10 meters of depth, the hydrostatic pressure (caused by the weight of the water column above you) will increase by approximately one atmosphere or one bar. The deepest trenches of the Pacific are almost 11 kilometers deep, implying that the pressure near their bottom amounts to 1.1 kilobar.

Apart from these record depths, most of the remaining oceans are deep enough to put an enormous pressure on the creatures that live down there. If you count all the water volume at more than 1,000 meters below the surface as belonging to the deep sea, this definition covers three-quarters of the oceans' total volume, or more than 60 percent of the total volume of the biosphere.

The oceans are, indeed, part of the biosphere down to and including their deepest trenches (and most probably the sediments of the ocean floor as well), as life has been found at all depths. The "dead zone" that the otherwise amazingly prescient writer Jules Verne (1828–1905) predicted to lie beyond 12 kilometers depth does not in fact exist. Whether it would exist if the oceans were actually as deep as Verne expected them to be remains open to idle speculation. Verne, however, in populating the seas down to 11 kilometers and having Captain Nemo wonder how organisms cope with the pressure, was much closer to the truth than the British oceanographer Edward Forbes, who, in the 1840s, strictly excluded the possibility of life beyond a depth of 600 meters.

The first samples of deep sea microorganisms were collected by the French research ships *Talisman* and *Travailleur* in 1882 and 1883 from depths of up to 5,100 meters. A. Certes, the biologist in charge of the analysis of these samples, had speculated about deep sea microbes before, on the grounds that none of the explorations of the sea floor using drag nets had ever brought any organic material (such as dead animals) to the surface. Such material, thought Certes, had to be degraded by microorganisms populating the sea floor. Therefore, he cultivated the samples of seawater and sea floor sediment brought in by the two research vessels with all the precautions of the microbiological practices that had only recently been introduced by Louis Pasteur.[5] For instance, he inoculated sterilized sea-water with a small sea floor sample and incubated it for several days. Of more than 100 samples incubated by Certes in the presence of oxygen (but under strict exclusion of contamination from the air), only four remained without detectable bacterial growth, as he reported to the Académie des Sciences in 1884.

Later that year, he reported additional experiments showing that the deep sea microbes remained metabolically active at pressures of up to 600 bar. In contrast, yeast, intestinal bacteria, and other nonadapted micro-organisms ceased to show any signs of life at pressures between 400 and 500 bar, as the physician and physiologist Paul Regnard (1850–1927) reported to the Académie in the same year. Regnard also found that yeast subjected to pressure obviously transforms into an inactive but resistant state, which upon release of pressure recovers within one hour and returns to normal growth behavior. Regnard also carried out similar experiments with higher organisms, which, however, showed higher sensitivity. While leeches and mollusks could be revived after pressure treatment of up to 600 bar, even 200 bar proved fatal for some fish species.

This promising start of the discipline of barobiology (the science of life under elevated pressure) was followed by a rather unproductive period lasting several decades, during which only sporadic and contradictory results became known. It was only after the work of Eyring[6] had led to a better understanding of the physicochemical foundations of the influence of pressure on chemical reactions that more systematic investigations were begun. In the 1940s, scientists realized that keeping other thermodynamic variables (such as temperature) constant while varying the pressure was of utmost importance—differences in these parameters had obviously led to the apparent contradictions between earlier studies.

The new, more systematic barobiology was most notably influenced by Claude E. ZoBell, who started in 1946 (working at Princeton University and later at the Scripps Institute of Oceanography of the University of California) to investigate the influence of high hydrostatic pressure both on marine microbes and on the common model organisms of cell biology such as the intestinal bacterium *Escherichia coli*. Even though high pressure biology and biochemistry have been studied systematically by a growing number of laboratories during the half century following the first papers by ZoBell, progress in this field has lagged behind comparable investigations into the effects of and adaptation to, say, high temperatures. This may be partly due to the "heavy" apparatus needed for high pressure studies, which may be frightening for students and researchers in biology.

Moreover, recovering samples from the deep sea is far from trivial. Although the aforementioned research submarine *Alvin* has been in service since 1964 and has been enormously useful for marine biologists and geologists alike, its maximum operating depth is 4,500 meters—less than half the way down into the trenches of the Pacific (see Sidelines, "On Diving"). Among the microorganisms recovered from the depths accessible to *Alvin*, none has been shown to thrive exclusively under elevated pressure, despite the efforts of the Woods Hole biologist Holger Jannasch, who developed a sampling device allowing the recovery of deep sea samples at constant *in situ* pressure. In principle, even creatures that need permanently high pressure to survive (obligate barophiles) should survive the transfer to the laboratory in this device.

The first obligate barophilic microorganism was described in 1981 by Aristide A. Yayanos, based at the Scripps Institute. His group recovered a dead amphipode (*Hirondellea gigas*) from a site in the Mariane Trench, 10,500 meters deep, which they then incubated in the laboratory at the *in situ* pressure of 1,050 bar. The bacterium isolated from this slightly unusual enrichment culture (and known by the somewhat unimaginative name MT 41) grows optimally at pressures between 300 and 700 bar and temperatures between 2 and 4°C.

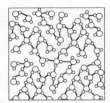 **sidelines** _____

On Diving

Sperm whales (*Physeter catodon*) are the world's deepest-diving mammals. These giants of up to 20 meters in length have been encountered at depths of up to 2,440 meters below the surface. This implies that their blood and every individual cell of their 60-ton body is subjected to a hydrostatic pressure of 244 bar.

Land-dwelling mammals like ourselves, however, find it extremely difficult to reach such depths. Without technical equipment we can hardly dive deeper than three or four meters, as we cannot hold our breath for more than a minute or two.

Diving suits supplied with fresh air from onboard a ship were already in use by the early 19th century. However, they tend to reduce the mobility of the diver and are still limited to a depth of about 440 meters. Diving in depths of up to 40 meters became an option when the scuba (short for "self-contained underwater breathing apparatus") was invented by the Frenchmen Jacques-Yves Cousteau (1910–1997, best known for his documentary films on marine life) and Emile Gagnan in 1943. Beyond 40 meters, however, the pressure on the diver's lungs becomes a serious problem. Furthermore, the requirement of surfacing slowly to avoid the phenomenon known as the bends limits the depth that can be reached. Under the elevated pressure the blood dissolves more gas (mainly nitrogen) than it normally would. If a diver rises too quickly, the gas forms bubbles and can cause embolism and tissue damage.

Thus, to explore the deep sea, humans need submarines. In the simplest version, this can be a rigid spherical or cigar-shaped container let down from a ship. In 1930, Otis Barton invented such a diving sphere (bathysphere) and used it to go to a record depth of 923 meters. Although this was a remarkable achievement at the time, it was only $\frac{1}{12}$ of the way down to the deep trenches.

More than two decades later, the Swiss engineer Auguste Piccard developed an independent, manned, and navigable deep sea vessel, the bathyscaph. This zeppelin of the deep sea consists of a cigar-shaped carrier unit filled with petrol. Beneath it, a sphere of two meters diameter surrounded by a steel wall nine centimeters thick serves as the cockpit. The prototype reached 3,150 meters in 1953. Using an improved version baptized *Trieste*, Piccard's son Jacques and the American navy officer Don Walsh achieved the oceanographic equivalent of the moon landing—in 1960 they reached the bottom of the Challenger Deep in the Mariana Trench, near Guam, 10,912 meters below the surface of the Pacific. In contrast to the moon landings, however, this pioneering achievement has not yet been repeated. The deepest-diving manned submarine in service today is the Japanese deep sea vessel *Shinkai 6500*, which can carry a crew of three down to 6,500 meters below sea level. Somewhat less costly and more versatile vessels can be employed in shallower waters. The best known and oldest workhorse of marine biology, the research submarine *Alvin*, serving the Woods Hole Oceanographic Institution since 1963, can dive down to 4,500 meters and thus cover large parts of the areas most interesting for biologists. Of course, the vast majority of submarines are not built for marine biologists, but for the navies of

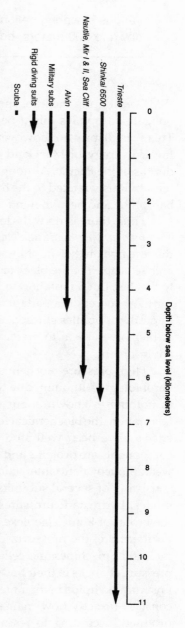

various seafaring nations. Military submarines are estimated to dive
down to 900 meters, but the exact details are classified.

Considering the idiosyncratic character of MT 41, Yayanos suggested
obligate barophiles might be a separate branch in the genealogy of life.
This was disproven, however, when Jody Deming, a marine biologist at
the University of Maryland, isolated a second such organism from the
depths of the Pacific. Although it was classified as a new genus (*Colwellia*),
it is obviously related to the bacterial genus *Vibrio*, which is generally non-
barophilic and best known for including the cholera germ, *Vibrio cholerae*.

Thus, there is no well-described standard barophile in the sense that
Thermotoga, Thermus, Sulfolobus, and *Pyrococcus* have become paradigms in
the research on thermophilic adaptation. Of course, all the aforementioned
archaebacteria of the black smoker biotopes live at *in situ* pressures of 100
to 500 bar, but adaptation to high temperatures seems to be more critical
for their success than adaptation to high pressures. All the black smoker
hyperthermophiles studied can still thrive at atmospheric pressure, some
of them even seem to prefer this condition to the pressure of their home
environment.

High pressure biochemistry has been arguably more successful than
barobiology. Although, due to the lack of suitable model organisms, next
to nothing is known about the molecular basis of (obligate) barophilic
adaptation, the biochemical foundations of the sensitivity against pressure
effects have been well studied since the 1970s. The most sensitive cell
components are proteins and complex molecular systems held together by
weak (noncovalent) interactions. Virus particles, ribosomes, and enzymes
consisting of several subunits can be separated into their components by
modest hydrostatic pressures (ca. 500 bar), as has been demonstrated by
the group of Rainer Jaenicke at the University of Regensburg and others.
Furthermore, the pressurized cell suffers from the loss of mobility of its
molecular machines and cell membrane. Certain fish can even respond to
pressure changes in their body fluids by regulating the ratio of saturated to
unsaturated lipids in their membranes so as to keep the membrane fluidity
constant (exactly how margarine producers control the softness of their
product), according to research reported by the group of Alister Macdonald at the University of Aberdeen.

A Light in the Dark:
Luminescent Creatures of the Deep Sea

Besides the cold, high pressure, and lack of nutrients, there is an additional stress factor hindering life in the deep sea: the total absence of light. The uppermost 200 meters of the water get some sunlight during daytime, but beyond that, and especially in the deep sea, there is absolute darkness 24 hours per day. Apart from the fact that this condition rules out photosynthesis as an energy source, it also makes life hard for animals that can see neither predators, prey, nor potential mates.

Surprisingly, though, nature seems to find it relatively easy to bring some light into the darkness. Evolution has invented bioluminescence— the conversion of biochemical energy into light—more than 30 times independently, as can be deduced from the sporadic distribution of glow-in-the-dark creatures among fish, insects, mushrooms, and bacteria. More than 30 different, unrelated enzymes, generically labeled luciferases,[7] can make as many different substrates, the luciferins, light up.

The concepts of luciferin and luciferase were introduced by the pioneer of bioluminescence research, the French physiologist Raphael Dubois, who compiled the first insights into bioluminescence and about the inverse reaction—photosynthesis—in his book, *La vie et la lumière* (*Light and Life*), published in 1914. In his classic experiment, he submerged the brightly shining luminescence organ of the Caribbean beetle *Pyrophorus* in boiling water, which he cooled quickly after the light went out. He crushed the luminescent organ of a second beetle with a little bit of water at room temperature until it stopped glowing. Finally, he combined the two liquids and obtained a vividly glowing solution. Dubois called the heat-stable compound in the hot-water extract *luciférin*, and the active component of the cold-water extract *luciférase*. In the course of further investigations, he found that luciferase catalyzes a reaction of luciferin with oxygen—a mechanism that still holds for all luciferases known today. Most of them were discovered by E. Newton Harvey, who published his first paper on bioluminescence in 1913 and remained faithful to this field for a lifetime.

Luminescence phenomena vary as widely as the organisms that produce them—ranging from a weak glow to the headlights of the flashlight fish (*Photoblepharon palpebratus*), the light from which reaches more than 30 meters through water. The purpose (i.e., the evolutionary advantage) of the light production can often be identified as serving defense, hunting, or communication with members of the same species. In some cases, the purpose is unknown or debated. There is a possibility that some luminescence phenomena are just unwanted by-products of a metabolic reaction favored by evolution for reasons completely independent of the light production.

An interesting example of how light can serve as camouflage and thus give protection from predators is provided by the hatchet fish (*Argyropelecus aculeatus*). The light organs at the underside of this small fish emit a faint bluish light matching the weak filtered sunlight perceived by predators coming from below. Thus it breaks the silhouette of the fish and makes it more difficult for predators to spot. In contrast, the flashlight fish uses its headlights, which are populated by symbiontic luminescent bacteria, to hunt prey in the dark. It does not live in the deep sea, but takes advantage of its strong light to specialize in night hunting.

Various species of fish, including several found in the deep sea, use luminescence organs as bait. For instance, the deep sea anglerfish (*Melanocoetus johnsoni*) (Figure 8) has a specialized fin carrying a small light like a lantern just above its upward-oriented mouth. Anglerfish actually possess the patience and the leisurely lifestyle of anglers—they are bad swimmers and hunters and spend most of their time lying on the sea floor waiting for prey, which they hope their glowing bait will attract.

Figure 8. The deep sea anglerfish (*Melanocoetus johnsoni*) attracts its prey with the help of a rod-like fin growing from above its upward-turned mouth. Like many light phenomena in other fish species, the yellow-green glow of the anglerfish's bait is produced by bacteria living in a symbiotic relation with the fish.

Obviously, our knowledge of the deep sea and its inhabitants is still very fragmentary. Thus, it is hard to estimate whether the dark areas of the oceans have actually produced more luminescent creatures than other biotopes. Alternative methods of orientation in the dark are via the sense of touch, or the exact analysis of fluctuations and shock waves in the water, or sonar as used by bats.

Surprising discoveries, such as the black smoker biotopes or, in earlier decades of our century, giant squid or the coelacanth (*Latimeria chalumnae*) believed to be extinct, keep reminding us that the deep sea may still hold many surprises and possibly answers to more general questions such as the origin of life.

Travel to the Center of the Earth: The Deep Subsurface as a Biotope

Deep underneath the surface of the Earth there are quite unusual creatures—their heads are so hard you couldn't harm them by beating them, but their feet, carrying one toe each, are soft and sensitive. These goblins are almost as wide as they are tall, sleep during the day and work at night, and keep herds of grotesque pets that look like a cat with a giraffe's legs, or like a unicorn with an elephant's trunk. If they become impertinent, you can scare them off by stamping on their feet or by singing aloud.

The Scottish writer George Macdonald (1824–1905) obviously had to use a lot of imagination to keep his 11 children amused. And they were rather spoiled in that respect—they had been the test audience whose enthusiastic reaction ensured the publication of the adventures that a girl named Alice had in Wonderland. Thus, it is not surprising that Macdonald— notwithstanding his education as a scientist—was not content with the real world's living creatures and added some amazing subterranean species when he wrote his children's classic, *The Princess and the Goblin*, published in 1872.

While the human imagination has populated the space below ground floor level with all sorts of creatures since the times of Orpheus, science remained skeptical about the potential living space beneath the upmost layer of soil used by plants and animals. The textbook wisdom said that all life on Earth depends—in some way or another—on the photosynthesis carried out by green plants. Some organisms digest the carbohydrates that the plants build from water, carbon dioxide, and solar energy, while others use the oxygen originating from the same process. Even in the dark parts of the oceans, creatures can feed on the biological material sinking down from the top layer, so they, too, live on solar energy. The dogma only started crumbling when the black smoker biotopes were discovered in the deep sea. However, the generally accepted statement that those are independent of the photosynthetic food chain mainly relies on a quantitative argument. The food supply from above is much too scarce to explain the population densities found in these biotopes.

On the continents, however, if one drills deep enough, the supply of organic nutrients not only becomes scarce, it becomes zero. And thus, scientists thought, there couldn't be *any* life in the deep subsurface. However, detailed investigations, often carried out as by-products of geological drilling projects, have shown that there is life underneath layers of 500 meters of solid rock. Some of the bacteria living down there might well be useful for us living on the surface high above. They might, for instance, be able to degrade toxic chemicals that could otherwise spoil our drinking water, or they could help to exploit oil reservoirs that could not be used otherwise. On the basis of such ideas, the U.S. Department of Energy launched a research program entitled "The Microbiology of the Deep Subsurface" in 1985. Its aim was to discover microorganisms at greater depths and to collect fundamental knowledge about these organisms, their ecology, and their possible role in the purification of groundwater.

In its first years the project surpassed all expectations. The water that the scientists pressed out of the porous rock by pumping argon gas down into the drill hole was very much alive with bacteria. The researchers were surprised not only by the abundance of microorganisms, but also by their species diversity and by the relative ease with which they could be grown in laboratory cultures. At all depths accessible to their drill, they found bacteria that were able to digest a variety of inorganic and organic substrates and turn over the elements carbon, nitrogen, sulfur, iron, manganese, and phosphorus. These findings are hard to reconcile with the traditional view that the deep subsurface is poor in nutrients and therefore cannot contain very many species.

A possible explanation of these findings—coupled with an even bigger surprise—was provided in 1995 by a group of scientists working for the Department of Energy who had been drilling up to 1,500 meters deep into the basalt near the Columbia River in the state of Washington. Chemical analysis of the water recovered from the deep basalt layers suggested that methanogenic bacteria (bacteria that convert carbon dioxide and hydrogen gas to water and methane) were thriving down there. This wasn't unusual per se—many methanogens get hydrogen from other organisms and are thus part of the normal food chain. But the water samples from the deep rock layers contained more hydrogen than could conceivably be explained by any biological source. A hydrogen explosion in a dump filled with basalt waste provided the leading scientists Todd Stevens and James McKinley of the Pacific Northwest Laboratory with the crucial inspiration: The hydrogen must have originated from the basalt rock.

They succeeded in demonstrating that a mixture of oxygen-free water and ground basalt actually produces hydrogen. Their hypothetical explanation of the phenomenon is based on the assumption that the iron compounds that become dissolved through the weathering of the rock react in an unusual way with water molecules, leading to the production of molecular hydrogen (Figure 9). Furthermore, when Stevens and McKinley sealed the groundwater with the bacteria and freshly ground basalt for various lengths of time ranging from 14 days to one year, they found that certain groups of bacteria can indeed thrive on stone and water for up to one year.

Of course this is only indirect evidence with respect to the question of what nutrients the bacteria use in their natural habitat, one mile beneath the surface of the earth. As yet, one cannot completely rule out the possibility that they have more conventional sources of hydrogen and/or energy down there. It would also be of interest to elucidate the chemical mechanisms that lead to the production of hydrogen from the mix of water and basalt in more detail. For instance, it remains to be seen whether the bacteria actively promote the weathering of the basalt and the hydrogen production, which they could do, for example, by secreting acids.

However, these bacteria remain an unusual finding in any case, and their importance certainly reaches beyond that of just another curious ecological niche. Scientists investigating the origins of life on Earth could study these bacteria to find out how the first bacteria could survive in an oxygen-free environment, before evolution invented photosynthesis and thus introduced atmospheric oxygen.

Moreover, those who believe that our neighbor planet Mars may have carried life in primeval times, when its atmosphere was denser and pro-

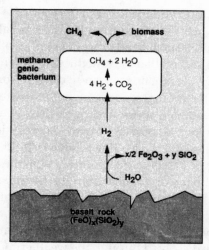

Figure 9. Hypothetical mechanism proposed for the formation of hydrogen from water and basalt rock. According to this view, the iron compounds of oxidation number II contained in the basalt reduce the hydrogen contained in water molecules to molecular hydrogen while being oxidized to iron III themselves. The poor solubility of the iron III compounds may pull the equilibrium in such a way as to facilitate this reaction, which would be rather unlikely to occur under normal conditions. The hydrogen, in turn, is required by methanogenic bacteria, which derive their metabolic energy from the formation of methane and water out of hydrogen and carbon dioxide.

vided a more agreeable climate and better protection from radiation, can take heart in this discovery. In the deep subsurface of the red planet, bacteria would be protected from the sun's ultraviolet light and might even survive on the stone-and-water diet to this day (see Chapter 6).

Less than one year after this discovery, "stone-eaters" were also discovered in a limestone cave. In contrast to the basalt bacteria, but similar to the sulfur oxidizers of the deep sea, the limestone bacteria seem to nourish a whole ecosystem. Scientists of the University of Cincinnati, Ohio, made this surprising discovery in a limestone cave in southern Romania that is partly flooded by relatively warm (21°C) groundwater containing hydrogen sulfide. The discoverers of the cave, the geologist Christian Lascu and the biology teacher Serban Sarbu, had already suspected something extraordinary when they first entered the cave in 1986. But only after the demise of the dictator Ceausescu in 1990 could Sarbu start a systematic investigation of the biotope with the help of the Cincinnati researchers.

Thirty terrestrial and 18 freshwater species were identified in the cave, more than half of which are exclusive to this location. Blind spiders, transparent crabs, water scorpions, and other unusual life forms make up this community, which originated five and a half million years ago and later was sealed so efficiently from the outside world that the oxygen used up was no longer replaced. The whole ecosystem receives no organic nutrients from outside. Wondering about the carbon source sustaining the biotope, the scientists investigated the bacterial mats that cover both the surface of the water and all the walls of the cave that lie above water level (Figure 10). By feeding the bacteria with isotope-labeled limestone, they could demonstrate that these are indeed capable of using the cave walls as their carbon source. Their energy is most probably derived from oxidation of the hydrogen sulfide dissolved in the groundwater. Although the details of their metabolism are not yet known, it has become obvious that they are independent of the photosynthetic food chain in a similar way to the black smoker bacteria. However, unlike the black smoker biotopes, the limestone ecosystem has not developed any symbiotic links. Differences

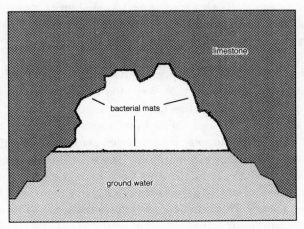

Figure 10. Schematic cross section of the bell-shaped air inclusions in the Movile cave. The mats on the water surface and on the cave walls consist of "stone-eating" bacteria. The air in the bubble contains less oxygen than normal air, and more carbon dioxide.

and common aspects of these ecosystems independent of the rest of the biosphere promise to yield interesting insights into the evolution of these small worlds.

Extra Dry: Survival in the Desert

When thinking about our planet as a whole, we tend to have an idealized vision of a paradise with blue oceans and continents covered with green vegetation. It is easy to forget that one-third of the continental surface is in fact covered by deserts. And this fraction may well be rising, if errors in the agricultural (mis)use of soil continue to be made. The largest desert areas are found in the subtropical areas lining the Tropic of Cancer and Tropic of Capricorn. Hot, dry air coming from the equatorial region determines the climates of the Sahara and the Mexican desert in the northern hemisphere as well as the Australian desert and the Kalahari in the southern hemisphere.

The Namib Desert stretches over 1,300 kilometers along the Atlantic coast of the African country Namibia, reaching into the southern parts of Angola. Traveling in its northern part in September 1859, the Austrian physician and botanist Friedrich M. J. Welwitsch (1806–1872) discovered a rather peculiar plant. Its short, turnip-shaped stem carried—apart from the two seed-leaves (cotyledons)—only two leaves, which were longer than one meter, touched the ground, and disintegrated from the tips backward. He sent some plants to London to have them examined at Kew Garden. They were found to represent not only a previously unknown species, but also a new family as well, the *Welwitschiaceae*. The Namib Desert's *Welwitschia mirabilis* has remained the only species.

Obviously, this plant can thrive in locations that receive less than 10 millimeters of rainfall per year. Even under extreme drought, it flowers every year and can live for a millennium. However, the thousands of seeds that each plant produces annually only have a fair chance in exceptionally good years, when relatively copious rainfall keeps the upper layer of the soil moist for a week. This time is sufficient for the seeds to germinate, and when the drought returns, the growth of their one vertical root has to keep up with the sinking of the upper limit of the moist soil. In this race, it can grow by up to 10 millimeters per day, and if it can keep up with the receding moisture until it reaches permanently moist layers,[8] the race is won and a new plant can thrive. As years with sufficient rainfall are few and far between, *Welwitschia* populations have a characteristic scattered age distribution. Only certain age groups are represented, but those in large numbers.

Plant physiologists are still puzzling over the question of how the plant can supply the leaves, which can measure up to two and a half meters in length and be up to 10 years old at the far end, with water without losing too much of it through transpiration on the way. Scientists at the University of Münster, Germany, have discovered regulatory mechanisms by which the plant can limit the loss of water through evaporation. They have also shown that, at the driest locations, the scarcity of water regulates the growth of the plant. Another interesting mechanism of adaptation was discovered when the scientists traced the pathways of certain chemical elements in the plant, notably of carbon, nitrogen, potassium, and magnesium. They found that *Welwitschia* can retrieve these elements from the dying ends of its leaves to be reused in the healthy parts of the plant. This unusual system of nutrient recycling allows the plant to survive for years without carbon intake, literally eating up its own leaves.

Lichens, too, are remarkably resistant against drought, as I mentioned earlier when discussing their cold-hardiness. Drought even helps them to withstand other stress factors; hence they are most frost-resistant when they are completely dried out. Among the bacteria, certain species of the genera *Bacillus* and *Clostridium* react to dry conditions by forming drought-resistant endospores, as will be shown in Chapter 3. One kind of bacterium, however, has evolved a different kind of drought resistance, which turned out to provide a rather surprising additional benefit.

The first representative of the genus *Deinococcus* was discovered in 1956 in a can of spoiled corned beef that had been sterilized with gamma rays. It was found that this method was no threat to *Deinococcus radiodurans*. The red-pigmented bacterium, which is harmless for humans, can withstand radioactivity and ultraviolet light at incredibly high doses. One thousand times the radiation density that would kill humans poses no problem to this bacterium.

The only ones to have a problem were the microbiologists, who at first couldn't explain this phenomenon. There is no place in the biosphere that is exposed to conditions that would require this resistance. Although the molecular mechanisms involved were studied in detail (they seem to have a particularly efficient way of repairing DNA, as I will explain in Chapter 3), the evolutionary advantage remained mysterious.

Only in 1996 did Valerie Mattimore and John Battista of Louisiana State University demonstrate that the radiation resistance is an accidental side effect of the equally remarkable drought resistance of the bacteria. Deinococci can be vacuum-dried, stored for six weeks in this state, and then revived. Mattimore and Battista analyzed more than 40 mutant bacterial strains that had lost the radiation resistance and found that all of them had also become drought-sensitive. Apparently, the oxygen from the air damages dried cells in a way that is similar to radiation damage and can be repaired by the same means.

Although radioactivity was not the selection criterion that favored this double resistance, radiation-hardiness doesn't have to be completely useless. Apart from the corned beef mentioned earlier, deinococci have also been found in the shielding bath of a cobalt 60 radiation source. They seem to be scattered over all continents and have been found in such unlikely places as weathered antarctic granite and the excrement of an elephant in a Japanese zoo, but a more detailed account of their preferred habitats is not yet possible.

Presumably, *Deinococcus* is not the only example of a naturally evolved stress resistance providing shelter from a different stress factor newly introduced by humans. For instance, halogenated hydrocarbons (those inert molecules used as cooling liquids and blamed for the decay of stratospheric ozone) do not occur naturally and have only been in existence for a couple of decades. Nevertheless, bacteria have been described that can degrade them, or even use them as the only carbon source. This phenomenon is most probably a similar side effect, rather than an incredibly fast adaptational reaction.

From the simple drought, the plain absence of water, we will now move on to conditions where water is physically present, but, alas, chemically not available.

Saturated with Salt:
The (Allegedly) Dead Sea as a Biotope

"Yam Hamelah" (Salt Lake) is the Hebrew name for the miracle of nature that we are used to calling—inappropriately, as we shall see—the

Dead Sea. Despite its literally biblical age, the world's saltiest lake—boasting a salt content of 28 percent—is very much alive. Some species of algae (genus *Dunaliella*) and various bacterial species tolerate the extreme salinity, which would be lethal for most other kinds of cells. The main inhabitants of the salty brine, however, are halobacteria—microbes from the domain of the archaebacteria (which we look at in more detail in Chapter 5), which have adapted to this environment so thoroughly that they cannot live at salt concentrations below 15 percent.

The Dead Sea is remarkable for a whole variety of other reasons as well. Its shores lie 400 meters below sea level and thus constitute the deepest site to be found on dry land on our planet. Furthermore, its surroundings are one of the hottest and driest regions of the Earth, as judged by year-average meteorological data. The valley that formed along the friction zone between two continental plates several million years ago and lost contact with the open sea one or two million years before our time practically serves as an evaporation trough for the waters of the river Jordan and some minor rivers and springs.

As the incoming freshwater is always lighter than the brine in the lake, a static layer structure has built up over centuries, with "fossil" saturated brine at the bottom (above sediments of solid salt precipitate) and newer, fresher waters further up. There is so little mixing that anaerobic microbes, which can only survive in the strict absence of oxygen, are found some 80 meters below the surface. However, since a major part of the waters normally feeding the lake have been diverted for irrigation purposes, the stability of the layer structure has been lost. Evaporation of freshwater from the surface that was not renewed rapidly enough led to the top becoming heavier than the bottom waters, and the lake literally turned upside down in the winter of 1978–1979, bringing deep fossil waters to the surface.

Of course, this was a major catastrophe for the microbes of the lake. In fact, the Dead Sea appeared to be properly dead in the months after the mixing. One year later, however, archaebacteria had begun to thrive again and covered the lake in their characteristic purple hue. Extreme halophiles seem to be absolutely unstoppable.

Scientists in Israel, the United States, Germany, and France are trying to crack the mysteries of why these "salt addicts" are so successful in this incredibly harsh environment. Although this is a classical example of fundamental—curiosity-driven—research, it has very early on yielded the discovery of one of the most promising materials derived from nature—the photosynthetic protein-pigment complex bacteriorhodopsin, which we will meet again in Chapter 4.

The main problem that the inhabitants of extremely salty waters have to face is that the salt has the tendency to keep hold of all the water molecules, so it can—by a process called osmosis—practically dry out any cells that do not react appropriately. Halotolerant bacteria and algae react by making large amounts of small organic molecules and accumulating them in the cell, to counteract the external osmotic pressure of the saline medium. Halophilic archaea, in contrast, cure one evil with an even worse evil. In order to keep the water in the cell, they accumulate salts in the cell to a concentration even slightly higher than the external salinity. Apart from other salts, their cytoplasm contains about four moles of potassium chloride per liter, which corresponds to eight times the total salinity of the oceans. (Note that a mole is an awfully big unit, accounting for some 6×10^{23} ion pairs or molecules. Thus, the world human population is currently approaching the numerical equivalent of 10^{-14} moles, or 10 femtomoles.)

This method, however, only moves the problem from the cellular level down to the molecular level. How cellular components, and specifically proteins, cope with high intracellular salt concentrations will be discussed in Chapter 3. We shall now deal with a different kind of stress, which is also caused by small ions.

Acid Heads and Basic Needs: Life at Extreme pH

Acids have been familiar to mankind for millennia. Vinegar, for instance, has been used for more than five thousand years both for conservation and seasoning purposes. When tasting something sour, we sense a physical parameter that is not as easily described as pressure or temperature. A few things had to happen before mankind could learn how to measure this parameter properly and understand it.

Pure sulfuric, hydrochloric, and nitric acids were first synthesized in Europe during the 13th century, presumably according to recipes of the Persian alchemist Al-Razi (850–923). The analytical chemistry developed in the 19th century introduced the titration of acids with bases (alkalimetry) or vice versa (acidimetry), but did not have very much use for this until organic chemistry supplied new indicator dyes and physical chemistry built a theoretical foundation to translate "acidity" into a well-defined and measurable quantity.

Both acidic and alkaline solutions are characterized by a change in the natural equilibrium between water molecules and two charged fragments into which they can be split, namely the hydrogen ion (H^+) and the hydroxide ion (OH^-). One liter of neutral water contains approximately 10^{-7} moles of each ion (along with some 55 moles of H_2O), and is therefore said to have a pH of 7.0, for the pH is defined as the negative exponent of the concentration of hydrogen ions. Acidic solutions contain an excess of hydrogen ions, so their pH is below 7, usually between 0 and 7, although there is no fundamental law ruling out pH values below 0. Alkaline solutions, in contrast, have less than the normal share of hydrogen ions, corresponding to pH values in the range of 7 to 14. Figure 11 gives some examples of typical pH values.

The pH is tremendously important for cells and their molecular machinery. If it is shifted by only one unit away from the physiological value (normally 7.5 to 8.5), virtually all of their functions are affected. Most biological macromolecules contain structural regions that react as acids or bases themselves, and therefore respond to changes in the pH of the environment by binding or releasing hydroxide or hydrogen ions. Such a reaction, of course, changes the charge carried by the site involved, which can in turn affect the structure and function of the whole macromolecule. Thus, it is not surprising that cells spend quite some effort on maintaining the internal pH at the physiological optimum, a task that can become quite difficult if the external pH differs from this value by several units.

Acid- and base-loving bacteria (acidophiles and alkalophiles) do not have very much in common, except that they are both extremophilic. They belong to different taxonomic groups, populate separate biotopes, and have different adaptation requirements. Hence we will deal with them separately, starting with the acidophiles.

We have already made the acquaintance of one genus of acidophilic bacteria. The hyperthermophilic archaea of the genus *Sulfolobus* found in hot springs and solfatara fields tend to be acidophilic in addition to having a thermophilic adaptation. Small surprise, as it is their own activity that makes their environment acidic: They convert hydrogen sulfide to sulfuric acid. The first representative of a group that is less thermophilic but more spectacularly acidophilic than *Sulfolobus* was isolated by Thomas Brock's laboratory from smoldering coal refuse in 1970 and was named *Thermo-*

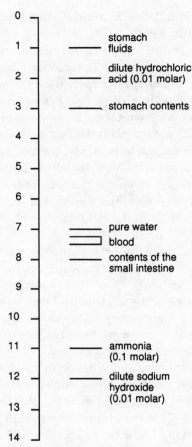

Figure 11. Typical pH values of some selected liquids.

plasma acidophilum. It grows optimally at 59°C and a pH just below 2.0, and it can just about get along at pH 0.4.

   Thermoplasma acidophilum held the record for acidophilic bacteria and archaea for little less than a quarter of a century, until the group of Wolfram Zillig, of the Max-Planck-Institut for Biochemistry at Martinsried (Bavaria), found two even more acidophilic species of a new genus in solfatara fields in Japan. The coccoid (ball-shaped) microbes *Picrophilus oshimae* and *Picrophilus torridus* thrive most happily at 60°C and pH 0.5, and they are still active at pH 0.0. If the environmental pH is shifted toward neutral conditions, however, the microbes stop growing at pH 3.5 and start to disintegrate at pH 5.0.

Although most of the extremophilic species identified so far belong to the domain of the archaebacteria, fungi are well able to compete as far as acidophily is concerned. In fact, the acidophilic fungus *Acontium velatum* was one of the very first extremophiles known to science. In 1943, R. L. Starkey and S. A. Waksman established that this species can thrive on a culture medium containing two and a half moles of sulfuric acid per liter (they do not quote the pH, but it must have been near zero). Equally amazing is the finding that the fungus can grow in dilute sulfuric acid saturated with copper sulfate, although copper ions are toxic for most kinds of cells. Only a few years later, other acidophilic fungi were described, and later they were joined by an algal species to fly the flag of eukaryotic extremists.

It is not clear how these organisms manage to keep the hydrogen ions out of their cells despite an external concentration seven orders of magnitude above the internal one. A little more is known about the exact opposite phenomenon, the adaptation of alkalophiles.

Most of the known alkalophiles, and among them the best-studied ones, belong to the genus *Bacillus*. *Bacillus alcalophilus*, for instance, is the standard model system for studies of adaptation to high pH. Alkalophilic bacilli are widespread and can even be found in rather unspectacular (not very basic) environments such as garden soil.

Bacillus alcalophilus thrives at pH values beyond 10.0, while maintaining a normal physiological pH of 8.6 in its cytoplasm. In the 1980s it was demonstrated that there is a membrane protein importing hydrogen ions (in exchange for sodium ions) into the cell to maintain the pH difference. As certain reactions relying on a positive difference between internal and external pH play an important role in the energy metabolism of normal (mesophilic) bacteria, alkalophiles are particularly interesting for researchers interested in bioenergetics, providing a unique opportunity to study alternatives to the classical mechanisms.

Generally speaking, however, chemical stress caused by high or low pH does not require as extensive an adaptational response as high pressure or temperature, as this chemical stress can be excluded from the cell. Adaptation is mainly required in the membrane proteins and ion channels, which deal with the import and export business of the cell. Of course, external organs like flagella have to withstand the external stress condition as well.

It is revealing to compare the pH extremists with the halophiles. Both have to deal with ion concentrations that differ drastically from the normal physiological conditions. The former deal with the problem on a cellular level, fighting it off at the membrane, while the latter have adapted the whole cell inventory to ion concentrations matching the conditions of the environment. We will revisit their adaptation mechanisms, which operate on the molecular rather than on the cellular scale, in more detail in Chapter 3.

Nature's Eco-Brigade: Oil-Degrading Bacteria

Tanker collisions or breakups spilling up to several hundred thousand tons of oil into the sea are not only remarkable for demonstrating the catastrophic consequences of human irresponsibility in dealing with our natural environment. They also provide spectacular examples of ecological chain reactions, which, under favorable circumstances, can lead to a surprisingly rapid self-healing of the affected ecosystem. In 1993, for instance, when the tanker *Braer* lost 85,000 tons of crude oil just off the Shetlands, experts feared the worst. After a few weeks, however, the pollution had miraculously disappeared.

As in many other miracles of nature, bacteria were involved in this one as well. Unfortunately, however, one cannot isolate a specific "anti-oil-spill" species of bacteria to be applied directly after an accident. Rather, it appears that a wide variety of bacterial species capable of eating up the various hydrocarbons that make up oil exist all over the globe. From this natural reservoir, multispecies communities develop rapidly as soon as an oil spill provides them with lots of hydrocarbons on which to feed. These communities seem to have specific task forces for certain kinds of molecules and develop in a way very much specific to the conditions of the given pollution event. If this bacterial oil removal did not exist, the North Sea would have been completely covered with a visible oil film.

Details of this teamwork seem to vary depending on the composition of the hydrocarbon mix and the environmental conditions, and they are not very well understood. One crucial aspect, however, has been clarified: the mechanism by which bacteria cope with the stress of a water-rejecting (hydrophobic) environment. They deal with this problem by making special, detergent-like molecules, the biotensids, which are at one end water-loving (hydrophilic) and hydrophobic at the other. Similarly, the building blocks of the cell membrane are hydrophobic on the inside of the membrane, but hydrophilic on both surfaces. Exposing a cell to a hydrophobic environment would disrupt the forces that keep the membrane together, as the hydrophobic parts of the membrane interior may find it attractive to turn outward to the hydrophobic environment. The membrane would disintegrate and form alternative structures, such as small spheres, with the hydrophilic parts inside. Using the biotensids as a protective shield on the outer surface of their membranes, bacteria can protect their membranes from being disrupted by the oily environment. Furthermore, they can send their homemade detergent molecules out to disperse the oil into microscopic droplets, which they can then attack and use as food.

Biological oil degradation is already used routinely in the regeneration of polluted soil following pipeline bursts or accidents in industrial plants. In particular, if the soil comes from a site with a long history of oil contamination, suitable microbes are already present, and only need a little encouragement such as aeration of the soil or feeding of essential trace elements. Regeneration stacks (Figure 12) can be piled up either directly at the site affected or in specialized soil regeneration plants.

The natural and biotechnical removal of oil spills is not the only connection between bacteria and oil. Current theories about the origin of

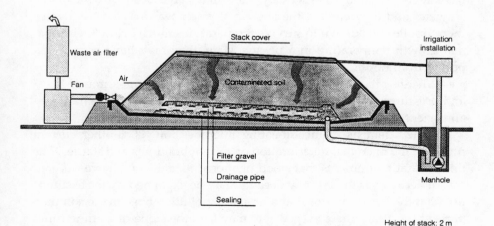

Height of stack: 2 m

Figure 12. Schematic representation of a regeneration stack, as used for the biological regeneration of oil-contaminated soil.

oil reservoirs rely on microbial involvement. However, no one has yet discovered oil in the making; hence we do not know anything about the microbes involved. (There is just one candidate, *Botryococcus braunii*, for a role in the initial steps in the formation of oil fields.) As the processes leading to the formation of oil most likely took place under high pressure, elevated temperatures, and in the absence of oxygen, these microbes would certainly deserve a place in this book.

While oil-making bacteria have hardly even bothered to prove their existence, oil-degrading bacteria can become a nuisance if they thrive in the wrong places, like fuel tanks or drilling platforms. In principle, their presence has to be feared anywhere, but they only become a problem when they find both hydrocarbons and additional nutrients such as oxygen and certain salts. This may, for instance, be the case in fuel tanks that are not cleaned properly.

In 1993, Karl Otto Stetter and his co-workers from the University of Regensburg managed to isolate hyperthermophilic bacteria from the production fluids obtained at four different oil fields. These microbes are most likely to be blamed for the generation of hydrogen sulfide gas, which is a common, but highly undesirable, phenomenon in oil production. It is not quite clear whether the bacteria had been present in the oil fields before the exploration and failed to degrade the oil due to the lack of additional nutrients. In this case, the contact with the salty seawater that is commonly used to press the oil out of the reservoirs underneath the sea floor would have revived them. Alternatively, they may have been present in the seawater and only entered the site when water was injected. As hyperthermophiles are known to survive temperatures much below their minimal growth temperature in a passive state, they may have been transported from hot biotopes by oceanic currents.

Wherever they came from, in oil fields with temperatures around 100°C and pressures around 400 bar, they seem to have found their paradise—and scientists will have to spend some effort to expel them from it.

Some scientists blame oil-eating microbes for yet another current problem: the microbial deterioration of ancient buildings and statues. The finding that monuments that stood unaltered for millennia have suffered severe decay over the last few decades may be explained by the assumption that the hydrocarbons that man-made pollution have made common parts of the atmosphere everywhere may feed the stone-degrading fungi and bacteria. The ruins on the Greek island of Delos, for instance, are infested with more than 80 different strains of black yeast. Scientists have managed to grow these microbes in the laboratory using a marble surface as a substrate and kerosene as the only additional food supply. The hydrocarbon hypothesis, first stated in 1966 by Wolfgang Krumbein at the University of Oldenburg in Germany, only slowly found acceptance. Nowadays, however, many scientists believe that the hydrocarbon content of the air causes more damage to ancient monuments than acid rain.

No matter whether the hypothesis is true or false, considering the enormous scale on which humanity uses hydrocarbons as fuels, allowing a considerable part to escape into the atmosphere unburnt, we should know more about how they interact with nature, and with microbes in particular. The reasoning, "oil and water don't mix; life needs water; therefore, there cannot be life in oil," has been proved false, but many aspects of this field remain to be explored.

Endnotes

1. The current definition for "hyperthermophilic" is that, under otherwise optimal growth conditions, the optimal temperature must be at least 80°C. Many hyperthermophiles cannot grow below 60°C, but survive temporary exposure to colder temperatures if they are carried, for instance, by drifts from one hot biotope to another.
2. Alfred Lothar Wegener (1880–1930), German geophysicist and meteorologist, led three expeditions to Greenland and died on the third.
3. Sima stands for silicon and magnesium compounds, the major components of the lower part of the earth's crust.
4. Other processes that contribute significantly to the material balance of the oceans include biomineralization (incorporation of inorganic materials in mollusk shells, bones, etc.) as well as the sedimentation of these biominerals after the death of their producers. If the oceans were nothing but a big evaporation pan for water supplied by rivers, they would have the salinity and composition of the Dead Sea.
5. See p. 100 and endnote 1 on p. 120 for details about Pasteur and his achievements.
6. The American physicochemist Henry Eyring (1901–1982) is best known for developing a theory of reaction kinetics named after him.
7. Nothing to do with the devil, but the word derives from Latin *lux* (light) and *ferre* (to bring), as does the name Lucifer.
8. Layers conducting groundwater (aquifers) exist even underneath the deserts. In places where they approach the surface or the water can move upward through cracks in the ground, an oasis can develop.

Updates

p. 32 New measurements reported in 2000 have shown that as the ice shield moves across the lake, some ice melts and mixes with the lake at the leading edge of the lake (the shore which meets the new ice first) while lake water refreezes at the opposite side. The drilling project has found traces of bacterial life in a layer of ice known to consist of such re-frozen lake water. Robotic instruments that can dig themselves through the ice without leaving an open hole that would permit contamination are being developed both for Lake Vostok and for future missions to Europa.

p. 38 I've reported how deep sperm whale can dive, but not explained how they manage to do it, as this was not known at the time of writing. It was only in April 2000 that a detailed investigation of the strategies of deep-diving mammals was published, based on observations from video cameras attached to freely diving animals. The material revealed that, while we would breathe in before diving, the animals empty their lungs and allow them to collapse at a certain depth. Thus they adjust their buoyancy in a way that allows them to go down using gravity rather than muscle force, which saves them enough energy to compensate for the lack of oxygen. Not taking any air on the dive also gets them past the bends problem.

p. 59 Hydrocarbon eaters of a different kind were discovered in the methane hydrate ("burning ice") deposits in the sea floor sediments. Methane hydrate is a solid formed by a combination of methane and water molecules at low temperatures and high pressures, and its occurrance is more widespread than anticipated. It is inhabited by bristle worms (polychaetes) and archaebacteria that can feed on the methane. Thus it forms a major extreme biotope, as well as a potential threat to our climate. If a significant part of the methane trapped in this state were set free, it could trigger runaway greenhouse warming.

3

The Cell's Survival Kit

When reading the previous chapter you may have wondered how cells can cope with all these kinds of environmental stress and extreme conditions. There is no general answer to this question. Depending on the kind of stress and on the evolutionary history of the organism involved, a wide variety of strategies can be observed. For instance, cells can stop a chemical stress factor at the cell wall, as do the organisms that thrive in acidic or alkaline solutions, the acido- and alkalophiles. Others, like the halophilic (salt-loving) bacteria, cope with it in their interior, on a molecular scale. Cells can counteract a burden by the massive synthesis of small molecules or by the increased production of specific proteins. Their reaction may be induced only in the short span following an emergency, or consist of a boost of a permanent part of the cell's metabolism. They can aim to prevent any damage, or guide the stress influence in such a way as to minimize damage, or allow the damage to occur and concentrate on efficient repair mechanisms. Higher (multicellular) organisms such as ourselves have additional levels of stress response, ranging from the physiological to the psychological (and, indeed, to the cultural/technical), which, however, are not the topic of this book. Finally, a species can brave harsh conditions on its own or in symbiosis with others.

This variety of cellular and biochemical responses to environmental stress can only be dealt with in selected examples. While I hope the various strategies presented in this chapter are representative, the list is by no means complete. After all, there is a real possibility that science does not yet know all the tricks in the cell's emergency repertoire.

Some of the mechanisms presented in this chapter (and particularly those in the Focus boxes) will contain biochemical details with which you may not be familiar. If you're keen to learn about them, you will find the necessary vocabulary in the glossary at the end of the book. If you're not quite so keen, please ignore the details. They will not be required for the chapters to follow, which will be less demanding than this one, promised.

The Heat Shock Response

Fruitflies are dreaded as a pest in marmalade and fruit juice factories, as they tend to deposit their eggs in overripened fruit. On the other hand, they are highly appreciated by geneticists and developmental biologists, who call them by their Latin name, *Drosophila*. The insect only needs eight days (at 25°C) for its complete development from the egg to the mature fly, thus facilitating the breeding of large numbers in the laboratory. Its importance as a workhorse for modern biology is only matched by the intestinal bacterium *Escherichia coli*.

Like many other research fields, the investigation of the heat shock response began with an observation made with fruitflies. In a paper published in 1962 and little noticed at the time, F. Ritossa described a kind of puff he had observed in the chromosomes from the salivary glands of *Drosophila busckii*. This new phenomenon could be induced by heat, nitrophenol, or sodium salicylate (a compound related to aspirin). During the following years, this stress response was mainly studied on a cellular level. It was found, for instance, that various other kinds of stress could induce the bulging, which occurs a few minutes after the onset of the stress condition, and that its appearance is coupled with an increased synthesis of RNA.

Investigation of the molecular details of this phenomenon only began in 1973, when A. Tissieres and H. K. Mitchell found that the occurrence of puffs is accompanied by the synthesis of a small number of new proteins. This group, the heat shock proteins, became a model system in which the mechanisms of gene regulation could be studied with great success (partly because their synthesis can be conveniently triggered "en bloc"). Most of the research was done on cell cultures derived from *Drosophila* species.

However, the heat shock response as a universal phenomenon only became fully appreciated after heat shock proteins were discovered in yeast in 1979, and thereafter—with the help of the newly developed two-dimensional electrophoresis methods described in Sidelines, "How to Hunt for Stress Proteins"—were found wherever researchers looked for them. Surprisingly, typical heat shock responses can be observed in psychrophilic algae, which tend to find a temperature of 5°C shocking, as well as in hyperthermophilic bacteria, whose growth optimum lies near 100°C. As the first information that electrophoresis provided about these newly discovered proteins was typically an approximate molecular weight, they often have names indicating their weight in kilodaltons, such as Hsp70 or Hsp104.

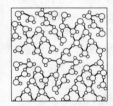 sidelines _____

How to Hunt for Stress Proteins

If biochemists want to know the approximate size of a biomolecule or if they want to sort a mixture of molecules by their masses, they normally use the method known as gel electrophoresis. In the most commonly used version, the protein is first boiled in a solution of the detergent SDS (sodium dodecyl sulfate), leading to protein molecules completely unfolded and wrapped in SDS molecules. As the SDS wrap provides a huge number of negative electrical charges, the charge of the protein chain itself becomes negligible. Thus, if an electric voltage drags it through a microporous molecular network (the gel), its mobility will depend only on the size of the molecule. Small proteins will migrate faster than larger ones.

Using this method, which is routinely applied in virtually every biochemistry laboratory in the world, one can neatly separate a mixture of up to a dozen proteins, provided their molecular weights are different. However, the method is almost useless for cell extracts with several thousand proteins. In order to separate the multitude of proteins of a whole cell, Patrick O'Farrell literally introduced a new dimension. The principle of the two-dimensional electrophoresis method that he developed is quite simple. You carry out an electrophoretic separation by one criterion, then turn the direction of the electric field by 90 degrees and·do a second separation, using a different, independent criterion. In the research done by O'Farrell and in most applications since, the first dimension is a separation by the content of basic and acidic groups. Protein molecules are forced to migrate in a gel containing a pH gradient (i.e., the acidity of the gel changes gradually from one end to the other). As the charges carried by basic and acidic groups depend on the pH, the net charge of each protein will be zero at a specific pH, the isoelectric point. When a protein in the gel has reached the position with this pH, it will stop moving. The second step, carried out at a right angle to the first, is an SDS electrophoresis as described earlier. Instead of spreading the different proteins over a linear distance of, say, 10 centimeters, you can spread them over a two-dimensional area, say 100 square centimeters. If the resolution of both methods is two millimeters, this gives you 50 points for the one-dimensional, but 2,500 points for the two-dimensional method.

Although it is much trickier to put into practice than the classical SDS electrophoresis, the two-dimensional method has found its place in the repertoire of many laboratories and was particularly useful for the basic investigations of the various kinds of stress response. The standard experiment to analyze the response of a microbial species to a stress factor X is quite simple in principle. Microbes are grown in parallel batches with and without X. The two-dimensional gels obtained from their cell extracts are compared, yielding information about the isoelectric point and molecular weight of the stress protein induced by X, enabling researchers to develop a purification protocol to isolate the protein from the cell extracts.

Many research groups around the world have carried out such comparative investigations, exposing countless organisms and cell lines to all conceivable kinds of stress, and very often they have observed changes of the protein patterns. One group that has been particularly successful in this research area is led by Frederick C. Neidhardt at the University of Michigan.

Some recently developed microanalytical methods have even improved upon the usefulness of two-dimensional electrophoresis. Methods developed by C. Eckerskorn and F. Lottspeich at the Max-Planck-Institute for Biochemistry in Martinsried near Munich, among others, allow scientists to use a spot from a gel (containing a billionth of a gram of a protein) to determine the amino acid composition and the beginning of the amino acid sequence of the protein. In most cases, these data will be sufficient to identify a protein unambiguously.

When comparing the protein patterns induced by heat shocks and other kinds of stress, scientists found overlaps, but not complete identity. Obviously, there is a set of universal stress proteins required in all kinds of emergency situations, while other proteins respond specifically to one kind of strain, such as heat, cold, lack of oxygen, or presence of detergents. Some shock proteins are always present at a basic low level and experience a manyfold increase in their synthesis rate during stress, while others are exclusively synthesized during the immediate shock response.

This was the state of knowledge in the mid-1980s: A whole bunch of shock-induced proteins had been identified, and the mechanisms of their genetic regulation were known in great detail. The only problem was, no one had the foggiest idea as to what the exact role of these proteins was in the cell and in which ways they can help the cell to cope with stress. This is not quite as astonishing as it looks at first glance, because many heat shock proteins carry out a type of function that had not yet been recognized to be necessary. Only in the late 1980s did it emerge that a new biological concept was required to understand the role of heat shock proteins—the concept of "molecular chaperones."

Heat Shock Proteins Acting as Molecular Chaperones

While *Drosophila* researchers were busy studying the cellular control of the heat shock response, other groups developed methods to study the folding of proteins in the reaction tube, and a third set of investigators addressed the question of which genes are required for the replication of bacteriophages (viruses that infect bacteria, also called "phages" for short). Nobody could have predicted that these three research areas would converge to the point where scientists from the three fields realized that they had been investigating different aspects of one phenomenon.

The American biochemist Christian B. Anfinsen (1916–1995) laid the foundations of the research field of protein folding with an experiment that required little more than a beaker. He was interested in the arrangement of the four cross-links (disulfide bridges) that stabilize the three-dimensional structure of the enzyme ribonuclease A. (He received the Nobel prize in chemistry for the elucidation of structure and function of this enzyme in 1972.) In order to open these bonds, he had to unfold the three-dimensional structure of the enzyme by adding the chemical compound urea—which is known to destabilize the folded structures of proteins—to a high concentration, so the disulfide bonds became accessible to the reagent used to reduce their sulfur atoms to the nonbinding S–H form. In this process, the protein lost all of its enzymatic activity but it was far from clear at the time whether this was due to the unfolding, the reduction of the disulfides, or both in combination. Therefore, Anfinsen tried to separate the effects in the reverse reaction. First he removed the denaturant by dialysis in the absence of oxygen—nothing happened. Then he allowed the oxygen of the air to be present when he removed the denaturant, and the activity of the enzyme returned as soon as the denaturant concentration had gone down. The oxygen had enabled formation of the disulfide bonds during the removal of the urea. Thus, for the first time in history, a denatured protein was renatured by human activity.

This kind of experiment, which has since been repeated millions of times for hundreds of proteins, demonstrates a fact that is much more important than the arrangement of disulfide bridges in ribonuclease A: It shows that the amino acid sequence of a protein contains the complete information required for the formation of its unique, biologically active three-dimensional structure. Although Anfinsen himself put forward the hypothesis that protein folding in the cell may be helped by other proteins, his experiment was interpreted as the ultimate evidence to the contrary. Proteins contain the complete structural information in their sequence, so why should other factors be involved in their folding?

One of the first observations indicating the involvement of helpers in structure formation was made in the area of bacteriophage research. When such a virus manages to inject its genetic material into a bacterial cell, the bacterium gets reprogrammed in such a way that its whole machinery of biomolecular synthesis switches to the production of as many copies of the bacteriophage (consisting of a rather short strand of nucleic acid and a self-assembling coat of protein subunits) as possible. Although viruses can be reconstituted *in vitro* from their individual components without requiring additional information or scaffolding, researchers were able to identify genes of the bacterial host that seemed to be essential for the correct assembly of new phages within an infected cell. As these genes affected the growth of bacteriophages and their absence could be partially compensated by mutations in the area of the phage genome labeled E, they were named *groE*. It turned out the genes coded for a large protein (GroEL) and a small one (GroES), a pair that rose to a fair degree of prominence in the years to follow.

The three research areas started to converge at the end of the 1980s, when it was found that GroEL and GroES are in fact identical with the bacterial heat shock proteins Hsp60 and Hsp10, respectively, and that they are involved in the folding of proteins and association of protein subunits in the cell. They were included in the newly discovered family of molecular chaperones.

The term "molecular chaperone" was first used by Ronald Laskey and his coworkers at the Medical Research Council (MRC) Laboratory of Molecular Biology (LMB) in Cambridge, England, in 1978. They observed that the protein nucleoplasmin helps the association of DNA with the DNA-binding histone proteins to form a correctly structured nucleosome by suppressing unwanted interactions—much as human chaperones in previous centuries were employed to stop daughters of upper class families from getting entangled in interactions considered inappropriate. In 1986, the plant biochemist R. John Ellis picked up this concept and used it to describe the function of a novel kind of protein that his group at the University of Warwick (England) had found to assist the folding and assembly of the plant protein rubisco. In the same year, Hugh Pelham of the LMB suggested that the heat shock proteins of the Hsp70 family might play a role in the assembly of other proteins. Three years later, Alexander Girshovich and his co-workers at the USSR Academy of Sciences in Pushchino demonstrated that GroEL binds transiently to newly synthesized, not yet folded polypeptide chains. Pierre Goloubinoff and George Lorimer found at DuPont in Wilmington, Delaware, that the GroE proteins assist the folding of proteins and association of protein subunits in the cell and use up adenosine triphosphate (ATP) in the process. Thus, a new biological principle had been discovered, and from then on, chaperone research started to spread like an epidemic.

And yet, Anfinsen was right—the sequence of a protein contains the complete information required for the formation of its three-dimensional structure. Therefore, many proteins can spontaneously form the correct structure when they are diluted from a denatured state into a suitable buffer. But Anfinsen was also right with his—seemingly contradictory—hypothesis: there are proteins that help others to fold correctly. The need for helping hands is a consequence of the high concentration of protein molecules in the cell. While the population density in an *in vitro* refolding experiment could be compared to Hyde Park on an early Sunday morning, proteins in the cell are crammed in a way strongly reminiscent of the London Tube during rush hour traffic. (If you don't believe me, look at the figures in David Goodsell's book, *The Machinery of Life*, published in 1993.) Under the crammed conditions in the cell—biophysicists actually speak of "crowding" effects—there are lots of undesired contacts between protein molecules, which can cause serious problems in folding and refolding.

The three-dimensional structure of proteins is held together by a large number of weak interactions, which, in contrast to the firm chemical links known as covalent bonds, are easy to make or break by a slight shift in the environmental conditions. These interactions include the aggregation tendency of water-avoiding (apolar, hydrophobic) amino acid side chains, which strive to evade the aqueous environment by clustering with others of their kind in a so-called hydrophobic core in the middle of the protein. As long as the protein chain is not yet folded, such potential binding sites are freely accessible and can easily interact with similar sites of different molecules, provided that there are any close by. And there are lots of potential binding sites in the overcrowded compartment that is the cell. It appears almost inevitable that this crowd of proteins, including many that are newly synthesized and not yet properly folded, should form non-specific intermolecular interactions rather than the structurally important intramolecular ones, eventually leading to large, insoluble aggregates of misfolded proteins falling out of the solution and becoming utterly useless for the cell. Considering this scenario, it appears very logical that molecular chaperones evolved to take care of the adolescent proteins and shield them from unwanted interactions. As an example, the current knowledge about one representative of this family, the bacterial heat shock protein and folding helper GroEL, is summarized in the Focus box, "Structure and Function of the Heat Shock Protein GroEL."

But what exactly is the role of chaperones in the heat shock response? Presumably the same as in normal, stress-free times, except that they have more jobs to do when high temperatures destabilize the structures of most proteins in the cell. Apart from the newly synthesized chains, they will also have to look after heat-denatured proteins and guide them back to their proper structures. For proteins that have suffered irreversible damage, they are thought also to be involved in the channeling toward the degradation pathways that ultimately lead to the recycling of the amino acid building blocks.

Antifreeze and Cold Shock Proteins

Liquid water is one of the most important requirements for life. Deep sea bacteria can only thrive at 110°C because the pressure of the water column increases the boiling point of water far beyond this temperature. While organisms at the upper limits of the biological temperature scale receive this kind of support from physical conditions, many inhabitants of ice-cold areas can actively keep water in the liquid state essential for their survival.

From everyday experience, we know various methods to prevent damage by freezing. Motorists, for instance, add antifreeze agents (such as the alcohol ethylene glycol) to the cooling and windscreen-wiping water of their vehicles; icy roads can be defrosted by salt. Both substances are very easily soluble in water, but cannot be included in ice crystals. Therefore, they pull the equilibrium between the liquid and the solid state of water in favor of the liquid and thus lower the freezing point by a few degrees.

Fish living in arctic waters cannot afford the luxury of a constantly high body temperature that we warm-blooded animals enjoy, but the motorist's method of mixing high concentrations of small molecules into their body liquids would get them into trouble as well. The natural tendency to equal concentrations (osmosis) would draw water into their cells and generate an osmotic pressure that could damage them at least as much as freezing and thawing. Therefore, various species of fish have evolved antifreeze proteins (AFP) that interact specifically with the crystallization

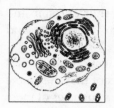 focus

Structure and Function of the Heat Shock Protein GroEL

The bacterial heat shock protein Hsp60, better known by the name GroEL, derived from the code first assigned to its gene, maintains a somewhat special position in the family of molecular chaperones. It operates at the end of the reaction path leading from the synthesis of the polypeptide chain on the ribosome (or from a protein unfolded by heating) via interactions with various other chaperones to the correctly folded protein. Therefore, its function can be more easily described than that of other chaperones, as we know the end product, the correctly folded native protein. And we know that the reaction uses up relatively large amounts of the cellular energy carrier ATP.

Investigations carried out by electron microscopy by Helen Saibil's group at Birkbeck College, London, and, more recently, various X-ray crystallographic studies beginning with the structure solved by the groups of Paul Sigler and Art Horwich at Yale University have shown that GroEL has the shape of a barrel composed of two rings of seven identical subunits. The smaller GroE protein, GroES, can bind to the open "top" of this barrel, closing it like a lid on a pot. Completely or partially unfolded proteins can be transiently bound by surfaces in the interior of the barrel structure. GroEL then runs through its functional cycle, which may or may not include the cooperation of the "lid" GroES and/or the consumption of ATP, depending on the specific requirements of individual substrate proteins, until the substrate protein is correctly folded. Native proteins are unable to bind to GroEL and are therefore released from the interior of the barrel.

The exact molecular details of the interaction between GroEL and its substrate proteins are still a matter of much controversy. One popular hypothesis states that the hollow barrel only acts as a cage, keeping the unfolded protein separate from others and thus suppressing intermolecular interactions. This model is known as the "Anfinsen cage" hypothesis, because it suggests that the cage provides the ideal condition for an Anfinsen style *in vitro* refolding experiment, namely "infinite dilution." Other scientists accord a more active role to GroEL, but without being able to agree on what that role exactly is. Some believe that it induces the unfolding of misfolded trapped species, which would otherwise never be able to find its way back to the correct path. Others prefer a more positive view that the cycles of substrate binding, release, and rebinding actively promote the correct folding.

A fundamental problem in this research area lies in the fact that the bound substrate is so disordered that it cannot be resolved in X-ray crystallographic structures. This challenge has led to the development of novel methodologies (for instance, the combined use of hydrogen exchange labeling and mass spectrometry, in which I was involved during my postdoctoral work with Sheena Radford at the Oxford Centre for Molecular Sciences), so there is a chance of more detailed knowledge emerging in the near future.

---

nuclei of ice, thus inhibiting the formation of ice and effectively lowering the freezing point in their immediate environment.

In contrast, certain species of frogs and turtles avoid damage by facilitating freezing of their body liquids using ice nucleation proteins. The rapid freezing simultaneously starting from many such nuclei keeps the ice crystals so small they cannot cause any mechanical damage. And even bacteria possess, in addition to the well-studied heat shock proteins, a set of cold shock proteins to help them buffer the consequences of a temperature drop.

So we are looking at specific interactions of biomolecules with ice nuclei that have only just started to form in order to either stop them from growing or facilitate their formation and growth—that sounds like a tricky problem requiring elaborate molecular structures to solve. Hence it came as a surprise when the first crystal structure of such a protein revealed a stunningly simple design. The antifreeze protein of the winter flounder (*Pseudopleuronectes americanus*) consists of only 37 amino acid residues and is completely wound up to form a single alpha helix with nine turns

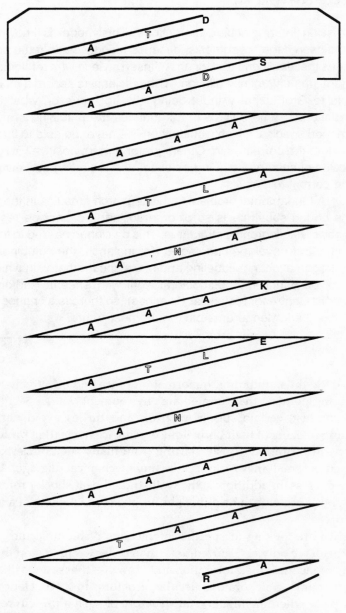

Figure 1. Schematic view of the helical structure of the winter flounder antifreeze protein. The amino acids represented by open letters are most probably involved in the binding of crystallization nuclei. Abbreviations: A, alanine; D, aspartic acid; E, glutamic acid; K, lysine; L, leucine; N, asparagine; R, arginine; S, serine; T, threonine.

(Figure 1). Researchers were surprised to find that such a simple structure lacking the water-excluding core that is thought to be important for protein structures can be stable and functional. They attribute the unusual stability of this single helix to special, cap-like structures linked up by hydrogen bonds at both ends of the screw.

It is equally amazing to note that the protein contains only 9 of the 20 different sorts of amino acid building blocks available. It is not very hard to guess which amino acid residues contribute to the ice-binding function, as alanines—unsuitable for this purpose—take more than half of the positions. Researchers believe that all 14 nonalanine residues have important roles either in helix stabilization or in the recognition and binding of ice crystals. The ice-binding motif was identified as a triad of the amino acids asparagine, threonine, and leucine, recurring in every third turn of the helix. The same regularity is also found in bigger helical AFPs with up to five repetitions of this motif. These residues form a surprisingly flat surface, from which the side chains of asparagine and threonine protrude only slightly. These can form hydrogen bonds with periodically recurring structures on certain surfaces of ice crystals, thus keeping the crystals from growing.

In contrast to the group of helical antifreeze proteins (type I AFPs) with their surprising simplicity, the type III AFPs found in other ocean fish, such as haddock, are somewhat more complex in their structures. Despite their relatively small size, they have a quite intricately folded structure with no obvious periodicity or any other hints of ice-binding motifs. After researchers arrive at the structure of a type III AFP by nuclear magnetic resonance (NMR) spectroscopy, they still need tedious investigations exchanging individual amino acids one by one to find which are essential for the interaction with ice crystals. Meanwhile, it appears that out of the eight sequence strands involved in formation of beta-pleated sheet structures, the one nearest to the end of the sequence (C-terminus) forms the binding site.

Researchers are still all at sea regarding the type II antifreeze proteins, which are found in herrings, for instance. There are no real structural data concerning these proteins, only hypothetical structural models based on a rather remote kinship with a group of plant proteins, the lectins.

Not very much more is known about the ice nucleation proteins, whose interaction with ice crystallization nuclei has the opposite effect, namely to facilitate nucleation and rapid freezing. This effect enables some species of frogs and turtles to survive freezing unharmed, even though up to 65 percent of their body's water content may turn into ice. Proteins with similar functions are secreted by certain microorganisms, specifically from the genera *Pseudomonas*, *Xanthomonas*, and *Erwinia*, in order to control ice formation in their immediate environment. As these proteins contain frequent repeats of certain sequence motifs, scientists suspect that a regular structure, possibly a large beta sheet, matches the periodicity of ice crystals.

Of course, the frogs will still find it stressful to be frozen, and there is some evidence that they cope with this by synthesizing specific cold shock proteins, in analogy to the heat shock proteins produced by most organisms upon exposure to higher temperatures.

Researchers hope to obtain more insight into the function of cold shock proteins by studying bacteria, which tend to be simpler in such fundamental things. Thus, the major cold shock protein of *E. coli*, CspA, was only identified and characterized in 1990, but its crystal structure was solved by 1994 (Figure 2). It is virtually identical with the structure of CspB from *Bacillus subtilis* determined one year earlier, indicating that this function is evolutionarily quite old and predates the split of the bacterial kingdom into Gram-positive and Gram-negative bacteria. Both contain structures typical for proteins binding to nucleic acids (there is a homology with the ribosomal protein S1, for instance). Although their function has not yet been elucidated, they are most probably not directly concerned

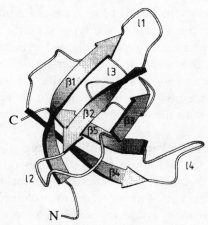

Figure 2. Schematic representation of the beta-sheet structure found in bacterial cold shock proteins such as CspA from *E. coli* or CspB from *B. subtilis*.

with the physical effects of the cold (as the AFPs are), but rather with helping the cell to adapt its functionality to the stress condition.

As a result of all these investigations, researchers not only hope to gain deeper insight into the mechanisms of response to extreme conditions. (In this field, the heat shock response has become a paradigm, while other extreme conditions such as cold, pressure, extreme pH, and high salt concentration are less well studied.) Biotechnologists also hope to translate the detailed knowledge of natural mechanisms into agricultural applications. They have already tried to get potato plants to express the winter flounder AFP, hoping to develop a potato variety able to thrive in the high Andes. An ice nucleation protein from *Pseudomonas syringae* already serves in the routine fabrication of artificial snow, and perhaps one day motorists will reach for a biological product to defrost their cars on cold winter mornings.

Adaptations by Changes of Amino Acid Sequences

The salt-loving archaebacteria of the Dead Sea protect themselves from the dehydrating effect of the high salt concentration in their environment by maintaining even higher salt concentrations in their cytoplasm, as discussed earlier. But this only shifts the problem from the cellular to the molecular level, as within the cell the salt will compete with proteins and other biomolecules for the essential solvent, water. Proteins from non-adapted species would not be functional at such high salt concentrations— they would aggregate to insoluble and useless lumps instead. Proteins from halophiles, in contrast, not only withstand these conditions, they even need salty brines to be fully functional. If one lowers the salt concentration starting from their "normal" physiological level (which is pretty close to saturation), their enzymatic activity will first decrease, then cease completely. Many halophilic proteins can actually be unfolded by withdrawal of salt to a concentration below 0.5 molar—a concentration that our tongues would still find unbearably salty. Ironically, acids, which can unfold normal proteins, can counteract these salt-withdrawal symptoms.

Scientists only began to understand all these peculiarities after crystal structures of two enzymes from the Dead Sea archaebacterium *Haloarcula marismortui* were solved by American–Israeli collaborative project teams. Both proteins, the metabolic enzyme malate dehydrogenase and the electron transfer protein ferredoxin, are wrapped in coats of acidic amino acid side chains[1] that in neutral media carry one negative charge each. This exceptionally high charge density is the reason why the protein surface can bind water molecules much more efficiently than that of a mesophilic protein and cannot be dehydrated in high salt concentrations. Compared with similar, nonhalophilic proteins, these enzymes can form up to 40 percent more hydrogen bonds with surrounding water molecules. In addition, the negative charges repel each other, an effect that can counteract aggregation, but can also disrupt the protein structure at lower salt concentrations, as a smaller number of positive ions is available to balance the negative charges. This is the reason why halophilic proteins can be inactivated and even unfolded by the removal of salt.

The third halophilic protein to have its molecular structure solved in detail was dihydrofolate reductase (DHFR). This enzyme catalyzes a reaction essential for the synthesis of new DNA and thus for cell division, and therefore it is an important target for cancer chemotherapy. That is why variants of this enzyme are well studied in organisms all the way from *Escherichia coli* to humans. This situation enabled Gerald Böhm and Rainer Jaenicke of the University of Regensburg as early as 1994 to predict the structure of halophilic DHFR from the known sequence of the amino acids and the expected similarity with the variants from nonhalophilic bacteria, whose crystal structures had been solved already. The predicted structure showed clusters of negatively charged amino acid residues in several places of the surface, while the site of substrate binding and catalytic activity carried most of the positive charges.

In 1997, Osnat Herzberg and her co-workers at the Center for Advanced Research in Biotechnology in Rockville, Maryland, finished determining the crystal structure of DHFR from *Haloferax volcanii*. While the experimental result confirmed the predicted overall pattern of a positively charged active site region surrounded by patches of negative charges, the detailed localization of some amino acids had to be corrected. As the proportion of matching amino acid residues in homologous proteins from different species can be as low as 30 percent, structure prediction tends to be a risky business.

In fact, the clustering of negative charges, which is held responsible for the instability of the protein structures at low salt concentration, turned out to be the only "typical halophilic" feature shared by all three halophilic proteins analyzed. The high number of salt bridges (electrostatic inter-

actions between oppositely charged residues) identified in malate dehydrogenase as a stabilizing feature was not found here, nor was the exceedingly negatively charged domain that seemed to play the role of an add-on
halophily module in ferredoxin. Thus, while the latter two structural
features are options that evolution may have used for adaptation to high
salinity in some cases but not in others, it remains true that clustered
negative charges (which destabilize the structure at low salt but favor its
water-solubility in high salt) are the most crucial factor for halophilic
adaptation.

Paradoxically, the mechanisms of molecular adaptation to high temperatures seem much less clear-cut, although a much larger body of structural data is available for proteins from thermophiles and hyperthermophiles. Whenever a research group believed it had found rules to predict
which amino acid exchange should improve the heat stability of a protein,
others found examples to the contrary. Only few general principles have
emerged:

- At room temperature, heat-stable proteins are more rigid than
  their nonthermophilic relatives, which can lead to a slowing down
  of enzymatic activity. Adaptation generally leads to homologous
  enzymes having comparable stabilities and flexibilities at "corresponding" temperatures, e.g., at the optimal growth temperatures
  for their respective organisms.
- In many cases, the bell curve describing the temperature dependence of stabilization energies is not shifted to higher temperatures, but is flattened (so the range of tolerated temperatures gets
  wider).
- Cavities in the packing of the protein can have a destabilizing
  effect. An optimized use of space (which can be realized experimentally by introduction of specific mutations) often improves the
  thermostability.
- As for the halophilic enzymes, charged amino acids play an important role for thermophilic enzymes as well. It is controversial,
  however, whether this has mainly to do with the binding of water
  molecules, or whether the attraction between opposite charges
  (so-called salt bridges) also plays a role.

Of course, a set of rules for thermostability would be enormously
useful for biotechnologists, for whom novel, heat-stable enzymes could

open a wide variety of new opportunities, as I will show in Chapter 4. But as yet, it is not quite as easy as that. There seem to be many ways to stabilize a protein against heat denaturation, and the best path to an optimal set of stability parameters has to be found for each protein individually.

Chemical Adaptation: Small Molecules

Even though small molecules are not suitable as cellular antifreeze agents, there are cases in which they can be used efficiently to cope with stress. For instance, the halotolerant and halophilic microbes mentioned earlier compensate for the external osmotic pressure by accumulating either organic molecules (in the case of *Dunaliella*) or salt, as mentioned before.

Apparently, small molecules can also be useful as a remedy for temperature-induced stress. The classic example is the disaccharide trehalose (Figure 3), which is similar to ordinary household sugar in that it consists of two ring-shaped sugar molecules linked together. While the two rings are glucose and fructose in ordinary sugar, trehalose is formed by two glucose rings. Initially, scientists believed that the yeast *Saccharomyces cerevisiae* only uses this carbohydrate as an energy store, but in 1989 the group of Andreas Wiemken at the University of Basel discovered evidence that trehalose can protect yeast from heat-induced damage.

Interestingly, yeast cells can "learn" a certain resistance against high temperatures if they are exposed to a brief heat shock before the longer heat treatment (five minutes at 52°C). The first heat shock stimulates the synthesis of trehalose, but of course it also induces heat shock proteins. The Swiss group, however, managed to find a stress condition that induces the production of all heat shock proteins, but not of trehalose. This treat-

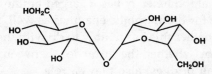

Figure 3. Chemical formula of the disaccharide trehalose, which is thought to have a protective effect against heat stress in yeast.

ment did not confer thermotolerance to the yeast cells, as heat shock and all other stress conditions that induce both heat shock protein and trehalose production had done.

While this was already strong evidence for the importance of trehalose in the acquired heat resistance of yeast, further studies from the Basel laboratory have also demonstrated this connection on a genetic level and have revealed that the presence of trehalose improves the heat stability of proteins in the cell.

Some New Tricks from the Cell's Repair Workshop

When an outside stress seriously damages a protein or a membrane component, the cell labels this molecule as waste and sends it off for recycling. Although proteins are expensive for the cell to produce, they can be replaced, and they are in fact replaced constantly. The situation is quite different for the genetic material, especially for germ line cells of higher organisms and for unicellular creatures. Any damage caused here not only affects the organism itself, but also its descendants. Therefore, it is enormously important that genetic damage can be repaired, and accordingly, nature has come up with a wide variety of repair mechanisms. An interesting example from recent research is how gene disruption induced by ultraviolet (UV) light can be repaired—using the energy of light, ironically.

Gene Repair Using Solar Energy

UV light can induce genetic mutations and kill cells (e.g., bacteria) efficiently. However, if bacteria are exposed to blue light immediately after a supposedly lethal dose of UV, this improves their chance of survival significantly. Although this mysterious phenomenon—known as photo-reactivation—was discovered in the 1930s, six decades had to pass before the underlying chemical mechanisms were understood.

The lethal effect of UV light can be understood as a chemical modification of DNA, which is made illegible by reactions triggered by this high-energy radiation. The most common modification is the reaction of two neighboring thymine bases leading to the formation of a four-membered carbon ring (cyclobutane ring) gluing the two bases together. This reaction, which is rather unusual by biochemical standards, blocks the DNA for those enzymes that normally read it to make DNA replicates or RNA transcripts.

On the molecular scale, the miracle of photoreactivation can be attributed to the activity of an enzyme named photolyase, which can bind to the damaged DNA, absorb blue light, and use its energy to split the cyclobutane ring and reestablish the two original thymine bases. For this purpose, a flavine nucleotide (i.e., a kind of coenzyme quite common in redox enzymes) bound to photolyase transfers a "borrowed" electron to the glued DNA bases, and gets it back after the successful split. Thus, the overall reaction is not an electron transfer (redox) reaction.

Photolyases can be found in creatures from all domains of life, in baker's yeast or goldfish, for instance, but they seem to be missing in a rather unpredictable fashion in some species, including that hairless primate whose activities have damaged the ozone layer to an extent that may well make UV-induced DNA damage a very serious problem in the future. The absence of photolyase does not mean that DNA damage cannot be repaired at all. In humans, for instance, the thymine dimer would be cut out of the strand and replaced by two new nucleotides by a repair mechanism much more complicated than the photolyase reaction.

The basic principles of the chemical mechanism of photoreactivation were solved in 1987, the crystal structure of photolyase in 1995, and some scientists thought the books could be closed on that case. However, a completely new chapter in photolyase research was opened in 1995 when a Japanese research group at the University of Kyoto cloned and characterized a gene that, in the fruitfly *Drosophila*, codes for another light-dependent repair enzyme, which was believed to have nothing to do with photolyase.

However, it emerged that this enzyme shares a common fundamental principle with the classical photolyases (see Focus, "The Growing Family of Photolyase Enzymes," for details). Furthermore, the new discovery revealed surprising genetic relationships, which also include a human gene that has not yet been characterized further.

The hope that insights from photolyase research may one day become useful in medical applications such as the treatment or avoidance of UV-induced skin cancer is one of the reasons for the continuing interest in photoreactivation. In the past, people with inherited oversensitivity and focus _____

The Growing Family of Photolyase Enzymes

The light-dependent repair enzyme that a Japanese research group discovered in *Drosophila* in 1992 removes a kind of cross-link occurring more rarely than the cyclobutane dimers mentioned in the main text. It deals with pairs of thymine bases in which the carbon atom number 4 of one base is linked to number 6 of the other, and is therefore called (6–4)-photolyase, to be distinguished from the classical, or CPD-photolyase (splitting cyclobutane–pyrimidine dimers). Due to the different probabilities of the two kinds of DNA damage, expression of recombinant (6–4)-photolyase in bacteria can be easily monitored when the bacteria contain a functional CPD-photolyase. Researchers simply had to increase the radiation dose to a level where even the more unlikely 6–4 cross-linking event became lethal. Bacterial colonies that could be photoreactivated even after this extra-harsh treatment obviously harbored a functional copy of the *Drosophila* (6–4)-photolyase gene in addition to their normal CPD-photolyase.

Surprisingly, this enzyme turned out to be a kind of "missing link." The *Drosophila* photolyase not only shares a remarkable sequence similarity with bacterial photolyases, but the Kyoto researchers could also identify a human gene that exhibits some homology with the *Drosophila* DNA (48 percent identity in the derived protein sequence) and might code for an as-yet-undiscovered (6–4)-photolyase. A further group of relatives turned up in plants: the family of blue-light photoreceptors, which have no connection with gene repair, but help to control plant development and growth as a function of light availability.

The reactions of all three protein families share some common principles. All of them absorb light in the blue to near-UV range with the help of a methenyltetrahydrofolate coenzyme (MHTF), which passes on the energy to a flavine nucleotide (FADH). The FADH then provides the electron that is needed for DNA repair or signal emission and gets it back at the end of the reaction.

Thus the deciphering of a single gene sequence has brought together three protein families not known to be related before. They all catch light in the same way, and then use it for their different purposes. More detailed comparative investigations of the three groups promise to yield interesting clues to the evolution of these unusual mechanisms.

However, the splitting of a cyclobutane ring with the help of a borrowed electron not only kept biologists busy, but also proved an interesting research field for organic chemists. In order to study the chemical mechanisms involved in more detail, Thomas Carell and his co-workers at the ETH Zürich developed a model compound that contained both a pyrimidine dimer (uracil instead of thymine) and an electron-donating flavin nucleotide within a relatively small molecule. As they had hoped, the researchers found that the model compound could be split by light when—and only when—the flavin nucleotide was in a reduced state (so it had electrons to spare).

The imitation of the photolyase reaction using small molecules in a reaction tube may one day become pharmacologically relevant. Apart from killing individual cells, UV damage can also lead to cancer. The best-studied example is the rare, inherited skin cancer xeroderma pigmentosum. The excision repair mechanism of human cells obviously fails in the very first step in these patients, thus leading to tumor growth. Perhaps photolyase research will one day teach us to help our cells repair their DNA and thus to prevent radiation-induced cancers.

---

very light-skinned people living in tropical countries were the main risk groups, but with the decrease of the ozone layer which protects us from the most devastating part of the sun's UV radiation, health risks from exposure to the Sun may well become a general problem. Lacking the elegant photolyase repair reaction, *Homo sapiens* will have to spend some of their *sapientia* on coping with this problem—and possibly find inspiration in the mechanisms of photoreactivation.

Rising from the Ashes:
The Miraculous Survival of *Deinococcus radiodurans*

In Chapter 2 we met those fanciful bacteria that obviously developed an incredible degree of resistance against radioactivity just as a side effect of their drought resistance. *Deinococcus radiodurans* even survives when radiation has turned its chromosome into a jigsaw puzzle of several hundred pieces. In less than 24 hours, the bacterium pieces the puzzle together without making a single mistake. It is not quite clear how this miracle works, but Michael Daly and Kenneth Minton, working at the National Institutes of Health, Bethesda, Maryland, have been investigating this phenomenon for years, and they think that the chromosome exists in two double-stranded copies linked by crossing over of strands, so-called Holliday junctions. This would mean that even in the worst-case scenario of both strands breaking or becoming illegible in the same region, there would always be an intact copy of the affected gene nearby from which the correct information could be read. Recombination of separate chromosomes as a mechanism of repair after stress-induced damage has also been proposed for *Sulfolobus acidocaldarius* (see Sidelines, "How Hot Love Helps Archaebacteria to Survive").

If this hypothesis is proved correct and found sufficient to explain the availability of the complete genetic information despite severe radiation damage, it will still be a bit of a puzzle to work out how *Deinococcus* reassembles the mess of fragmented crossed-over chromosomes into one intact copy. One gene (*recA*) essential for this recovery has been identified, but researchers will certainly carry on scratching their heads for a little while.

How to Cope with Defective Messengers

Information losses that eventually impair the biological function of a cell or organism can arise from other causes besides damage to DNA. The path leading from genes to proteins is sensitive to random reading errors, mechanical strand breaking, and chemical modifications induced by environmental factors in all its steps. Not all of the possible mechanisms of damage nor of repair or damage limitation by the cell are known, and only recently (1996) a process occurring on damaged messenger RNA (mRNA)

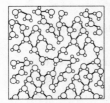 sidelines _____

How Hot Love Helps Archaebacteria to Survive

Even though the *chagrin d'amour* may last forever, there is still a major biological advantage to be gained from using sexual reproduction. In comparison with simple cell division accompanied by the odd random mutation, the mixing of the parents' genes allows for a greater genetic variation combined with a better stability of the collective gene pool of the population. In a sense, this is easier for us higher organisms, as our genetic material is organized in many pairs of chromosomes, so we just have to make sure we get one from each parent. Which of the two copies we receive, that is the lottery of life. The sperm that wins the race against millions of others may carry Granny's musicality or Grandad's big ears, and these genetic dispositions may override the corresponding genes from the maternal side or may lose out and remain silent for a generation—that's the name of the game. For bacteria, such a gene lottery is far more difficult to organize, as their genome normally comes as one double strand of DNA. Gene transfer only works if short DNA fragments can be cut out of the genome and then relocated with the help of viruses or mobile DNA rings called plasmids. Hence, for many microbial species, including the hyperthermophilic archaebacteria, sex used to be taboo.

In 1996, however, Dennis Grogan of the University of Cincinnati found the right kind of temptation to bring the heat- and acid-loving microbes of the species *Sulfolobus acidocaldarius* into intimate contact. He grew various mutants of the archaebacterium that, apart from the basic nutrients provided in the minimal medium, each needed one amino acid or coenzyme as a food additive to thrive. For instance, a mutant strain that lost the ability to produce the amino acid histidine from scratch would need a histidine supply in the growth medium. Then he mixed two of these mutant strains and incubated them on a medium containing neither of the required additives—it mainly consisted of the amino acid glutamic acid and of dilute sulfuric acid. Only through a gene transfer from one strain to the other could a new strain arise that would be able to thrive on a medium with no additives. In fact, Grogan found such "cured" colonies quite frequently, while a control experiment where the strains were not mixed but incubated separately on minimal medium only very rarely led to reverse mutations reestablishing the additive-independent wild type. The gene transfer was obviously active at temperatures of up to 84°C.

This was the first time that an exchange of genetic material between bacterial cells was demonstrated at such high temperatures, and also a first for thermophilic archaebacteria, and it opens up interesting opportunities and poses new questions. Grogan believes that *Sulfolobus* needs the transfer as a repair mechanism in case its DNA suffers heat-induced damage. In a similar way, primitive microbes in the early days of life could have coped with genetic defects caused by the UV irradiation, which was then much stronger than it is now. Since those times, archaebacteria are believed to have changed less than all other groups of organisms, thus their habits of living and loving may well be a window into the past (see Chapter 5).

---

and discovered by sheer serendipity provided some major surprises. If the mRNA lacks a stop signal (for instance, because the strand was broken before the stop), a piece of nucleic acid combining properties of both messenger and transfer RNA (tRNA) steps in and allows the synthesis to run to completion. But the freshly synthesized protein receives its death certificate at birth. The protein sequence coded by the newly discovered hybrid RNA is a degradation signal leading the defective protein directly to the cell's destruction machinery. With a detective's flair and a bit of luck, researchers have been able to decipher this emergency mechanism, which is unusual in a number of ways (see Focus, "The Case of the Missing Alanine").

Waiting for Better Times:
Sporulation as a Survival Strategy

After the molecular stress response, we will now look at the cellular ones, those in which the whole cell changes its properties in order to cope

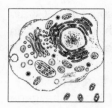 focus _____

The Case of the Missing Alanine—a Biochemical Detective Story

By the early 1990s, scientists believed they knew all the molecules involved in the protein biosynthesis mechanisms of *Escherichia coli*. Structural investigations of the ribosome and the associated factors progressed slowly but steadily and were considered unlikely to yield major surprises. Models of the overall functional cycles were well established. But then, a smallish RNA molecule emerged from nowhere and broke all the rules . . .

Richard J. Simpson and his co-workers at the Walter and Eliza Hall Institute of Medical Research in Parkville, Australia, were the first researchers to catch a glimpse of something strange. They tried to express a mouse protein, interleukin-6 (IL-6), in *E. coli*. The protein molecules obtained had different molecular weights, all of them too small, and when the researchers checked their sequences of both termini, they obtained the paradoxical result that the sequences of different length had identical beginnings and ends. It turned out that the correct IL-6 sequence ended prematurely at various positions and the unnatural end had been coupled to an alien peptide of 11 amino acids.

The last 10 of the 11 amino acids are coded in an *E. coli* gene named *ssrA*. Mutants that do not possess this gene cannot attach the 11-residue peptide either. Robert T. Sauer and his group at MIT suspected that this peptide might be a degradation signal—similar marker peptides were already known, including the intensely studied (but much bigger) ubiquitin. To test this hypothesis, they altered the gene sequences of two *E. coli* proteins in such a way that the 11-residue sequence became their natural C-terminus. In fact, both proteins were degraded within five minutes, while similar constructs differing from the signal sequence only in the last two positions survived more than one hour on average.

But what role does the *ssrA* gene play in this marking process? Transcription of this gene yields a strange RNA molecule. Its two ends combine to form a structure strongly reminiscent of the acceptor stem of a transfer RNA, and it can indeed be charged with the amino acid alanine—implying that the structural similarity must have fooled a specialized enzyme, the tRNA synthetase specific for alanine. And this finding matches nicely with the fact that the only amino acid of the 11-residue signal sequence that is *not* coded in the middle part of the *ssrA* RNA (which is obviously to be read as a messenger RNA despite the fact that the flanking regions can carry out tRNA functions) is— guess what—an alanine. Thus, the missing alanine was the key evidence that, in combination with a fair amount of Sherlock-Holmes-style criminalistic deductions, allowed Sauer and co-workers to put forward a hypothesis, test it experimentally, and finally track down the killer of the truncated proteins.

For the biosynthesis of a protein to be finished properly, the mRNA must contain a stop codon that will be recognized by release factors. These factors will make sure the synthesis is terminated and the ribosome released from the mRNA so both can be reused. If, however, the ribosome reaches the end of an RNA molecule without having encountered a stop signal (which may happen if the RNA strand suffers mechanical breaking at a random site), the whole system gets blocked. In order to prevent dozens of ribosomes from queuing up behind and becoming useless for the cell, the *ssrA* RNA starts a rescue operation. The tRNA-like part of the molecule charged with an alanine residue occupies the A (acceptor) site of the ribosome and tricks it into incorporating the alanine into the growing polypeptide, although there is no mRNA codon present in the A site that would suggest any such thing. After translocation, the pseudo-tRNA with the polypeptide including one noncoded alanine is in the P (peptidyl) site, and the middle region of the *ssrA* RNA now starts acting as an mRNA. The first codon (GCA) stands for another alanine, which this time comes in by the conventional way, via a normal alanyl-tRNA, and gets bound to the polypeptide. After the tenth codon, the *ssrA* RNA contains a stop signal (UAA), which, in interaction with the release factors, triggers a normal termination of the biosynthesis. The final protein product contains, after the site of the strand break, the treacherous 11-residue peptide that acts as a degradation signal in *E. coli*.

In order to test this hypothetical mechanism, which presumably only becomes active in the cell in the rare event of a broken mRNA, Sauer and co-workers provoked such a situation deliberately and on a massive scale by cloning a transcription stop into a protein gene before the (translation) stop codon, so all the mRNAs made from this particular gene in *E. coli* would have an "open end." In fact, the protein product of this construct was degraded within a few minutes, provided the cells possessed an intact *ssrA* gene. If the gene was inactivated, the protein did not carry the signal peptide and was not degraded very rapidly. In cells with a functional *ssrA* gene but lacking one of the enzymes essential for degradation, the protein could be found with the 11-residue peptide attached.

Thus, Sauer's group could prove unambiguously that in the case of an "open-ended" mRNA the *ssrA* RNA induces the attachment of the degradation signal. As an alternative to the mechanism outlined here, it would have still been conceivable that the attachment happened not during protein biosynthesis (cotranslationally), but rather on the mRNA level, or by chemical linkage of the ready-made peptides—the latter being the case for all other degradation signals known to date, including ubiquitin. Here, the detectives had to rely on circumstantial evidence, with the missing alanine in the key role. The combination of the findings that the first alanine of the 11-residue peptide is *not* coded in the sequence of the *ssrA* RNA, but that on the other hand the tRNA-like part of *ssrA* RNA can be charged with alanine—which only makes sense if this alanine gets smuggled into the sequence during protein biosynthesis—excludes both RNA splicing and polypeptide coupling with certainty. And thus the cotranslational mechanism described remains the only possibility.

This conclusion is somewhat hard to swallow for molecular biologists, as it implies that several sections of their textbooks need to be rewritten. First, this is the first example of a case in which a ribosome introduces an amino acid that is *not* coded on an mRNA into a growing polypeptide chain. Normally, the occupation of the A site with a tRNA carrying the amino acid to be incorporated into the polypeptide undergoes an exceedingly strict control of the correct codon–anticodon interaction. (Reading errors are quite rare and are more likely to consist in a shift of the reading frame than in a codon misreading.) Second, this is the first demonstration that the sequence information for one polypeptide can come from different mRNAs, so the ribosome can jump from one RNA to another. Although there were examples of ribosomes skipping short sequences within one mRNA, the finding of this jump from one mRNA to another came as a big surprise. Third, it had not been known that there are natural substances that can fool tRNA synthetases. Researchers have spent much effort on designing "minimal" tRNAs without suspecting that similar things existed in nature. Finally, the *ssrA* RNA is a completely new kind of biomolecule in that it functions as a tRNA in one part and as an mRNA in another. Various researchers have suggested calling it a tmRNA.

This last-mentioned discovery could also encourage new models for the evolution of protein biosynthesis. When the (hypothetical) RNA-world (see Chapter 5) diversified into the modern DNA–RNA–protein machinery, adapter molecules must have developed that had to precede the modern tRNAs in coupling between the RNA and amino acids. Perhaps the tmRNA is a leftover of these early adapter molecules—this might explain its unique properties.

---

with a stress condition. One of the most spectacular reactions of this kind is the formation of endospores by bacteria. Various species of the genera *Bacillus* and *Clostridium* respond to a scarcity of nutrients by a special mechanism of asymmetric cell division (Figure 4) leading to the formation of more resistant but inactive spores. Like plant seeds, endospores remain dormant until a signal from the outside world suggests conditions are favorable for germination.

In comparison with normal cells, the mature spore has quite a few distinguishing properties:

- It does not have a metabolism and its enzymes are inactive, so it cannot repair gene damage.
- Its chromosome is much more compact. The DNA binds special acid-soluble proteins, which are only synthesized during sporulation. The spore contains five to ten times less water per dry weight than an active cell.
- The pH of the cytoplasm is lower by one unit.

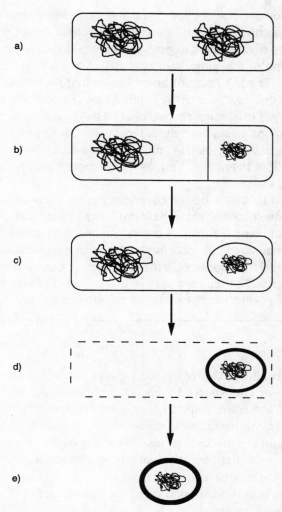

Figure 4. Sporulation in *Bacillus*. (a) The deviation from the pathway of normal cell division only begins after the chromosome is already duplicated. (b) A newly formed membrane divides the cell into two compartments of different sizes. The bigger one is called the spore mother cell, the smaller one the spore protoplast, which will develop into the spore as such. Each has one copy of the chromosome, and the copy in the protoplast becomes more condensed than in the mother cell. (c) The cytoplasm of the mother cell engulfs the protoplast, which develops into the prespore. (d) A cell wall of several layers is built on the outside of the prespore cell membrane. (e) The mother cell disintegrates and sets the fully formed spore free.

- The interior of the spore (the protoplast) contains a high concentration of the small molecule dipicolinic acid (DPA; Figure 5), which cannot be detected in normal bacterial cells.
- The cell wall is significantly less permeable.

While some of these measures increase the chance of survival, others compensate for disadvantages arising from the lack of an active metabolism. For instance, the cell wall must be sealed against the outside world more efficiently if there aren't any active enzymes that could fight off toxic substances invading the cell. As far as survival is concerned, the sporulation strategy is enormously successful:

- Reliable data prove that spores can survive unharmed for several decades. In 1995, Raúl Cano and Monica Borucki at the California Polytechnic State University at San Luis Obispo even managed to isolate bacterial spores from the guts of bees trapped in amber and revive them after a sleep that presumably lasted some 40 million years. In contrast to similar claims made earlier but remaining highly controversial, this result was so well documented and watertight that the renowned magazine *Science* published the research article.
- Heating to 85°C for 30 minutes kills exponentially growing cultures of *Bacillus subtilis* very efficiently; of its spores, however, 80 percent survive this treatment. Therefore, and because certain pathogenic germs such as *Clostridium tetani* are known to form spores, heat sterilization (in food processing or medical contexts) has to be carried out at 120°C at least.

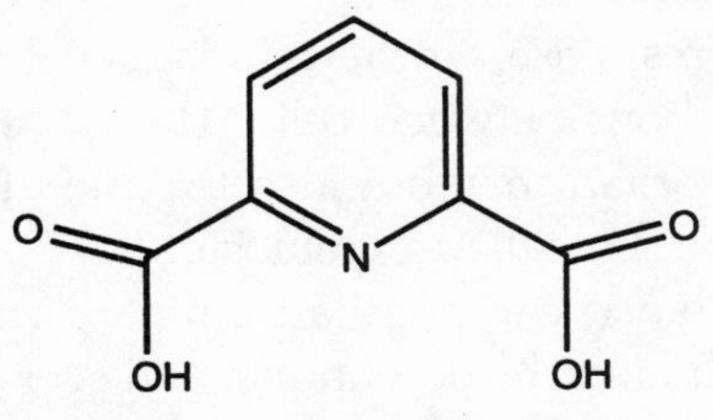

Figure 5. Chemical formula of dipicolinic acid (DPA), a substance found in high concentrations in bacterial spores.

- After one round of freeze-drying, the survivors count is 2 percent for active bacteria, but 100 percent for spores. Even after six rounds of freeze-drying and rehydration of spores, no damage could be detected.
- An oxidation treatment with hydrogen peroxide (20 minutes in a four-molar solution—quite enough to turn your hair white and peel your skin off) leaves 64 percent of *Bacillus subtilis* spores unharmed, while only one active bacterium in a million survives.

In sporulation we find a whole set of stress responses, some of which we have already encountered in other contexts, combined in an emergency program. Researchers have paid particular attention to the protection of the spore's DNA from the inevitable attack of chemicals, radiation, and other kinds of environmental stress factors. Apart from the protective effect of the very special packaging of the DNA in the spore, there is also a special "quick repair" program carried out during germination, when the spore starts to convert back into a normal, active bacterium.

Developmental biologists are also interested in sporulation as the simplest conceivable model for an enormously important process in the embryonic development of higher organisms: the division of a cell into two different, specialized cells. Crucial steps in both processes, in the development of an embryo as well as in sporulation, are controlled by proteins that regulate transcription (i.e., information transfer from gene to RNA), the so-called sigma factors. It turns out that sporulation is extremely useful, not only for bacteria, but for biologists as well.

Two's Company: Symbiosis Helps Species to Spread in Hostile Environments

From the stress response on the cellular level we now go one step further to the interactions between cells. These are not only crucial for higher, multicellular organisms, but also for a kind of close community between different species called symbiosis. It is often found that two species living in symbiosis can populate a biotope that would be unsuitable for either individually. The best known and economically most important example may be the fixation of nitrogen from the air by the bacteria colonizing the roots of certain plants, like the potato or the legumes, for instance. In the following, we shall deal with two (more extreme) examples in detail: the well-known and almost omnipresent lichens and the more exotic, but within their specific biotopes equally successful, tube worms of the deep sea.

Lichens

Lichens, well known as scaly coverings, some plain, some colorful, on old buildings, trees, and even on naked rock, can be found everywhere, from the arctic to the antarctic. They are obviously able to live on nothing but air, sunlight, and rain and cope with extreme changes of temperature at the same time. They are especially resistant against dry cold. At subzero temperatures, lichens can survive years of drought, and while they are dried out, they don't mind being cooled to $-200°C$. On the other hand, they also live happily on sun-facing roof tiles, where they may be heated to 70 to $80°C$.

The secret behind the success of this life form is that it really consists of two life forms. Close inspection reveals that the plantlike structures of lichens are built up by a symbiotic community of a fungus and a microbe that can perform photosynthesis (either an alga or a cyanobacterium), nested between the filaments of the fungus. The fungus receives nutrients from the alga, which can thrive on the energy from the sunlight and trace minerals from rain. The advantage for the alga may be that the fungus protects it from drought.

Despite their global spread and the interesting evolutionary questions arising from such symbiotic life forms, lichen research was neglected for a long time, until environmentalists began to appreciate lichens as indicators for air quality in the late 1970s. It is thought that every fifth fungus species can form a lichen, but some textbooks of mycology only mention lichens as a marginal curiosity, and many scientists still seem to think of lichens as an unimportant minority group in the tree of life.

In 1995, a group at the National Museum of Natural History in Washington, D.C., proved them wrong, using DNA analysis of 75 fungal species including 10 that form lichens. The researchers found out that the lichen-forming fungi are nothing like a separate group. In fact, the 10 species investigated belonged to five different groups that are more closely related to nonlichenogenic fungi than to each other.

This research shows that the symbiosis between fungi and photo-synthetic microbes has been "invented" several times during evolution. Far from being a specialty, it is a potential found in most branches of the fungal family tree.

Tube Worms and Sulfur Bacteria

A different kind of symbiosis, which also facilitates survival in extreme conditions, is found in the biotopes around black smokers in the deep sea. Interestingly, the community of the tube worm *Riftia pachyptila* with sulfur bacteria differs in almost every aspect from the lichen system.

While lichenogenic fungi and algae can also live separately, and the algal species may even be more widespread on its own, the tube worm is completely dependent on the nutrient supply from the sulfur bacteria. As these live within the cells of their host, scientists call the relationship an endosymbiosis. In contrast to the case of the lichens, where two primitive life forms combine to form highly complex, plantlike structures, we find here a higher (though still invertebrate) organism as a host for millions of primitive guest microbes.

The fact that tube worms are completely dependent on being fed by their guest microbes is mirrored in their body plan (Figure 6), which is totally lacking any organs that could be involved in food intake or processing. The worm has neither mouth, nor stomach, nor guts. Even though the waters around it are thick with bacteria, it cannot make use of this potential food source.

In order to unravel the secrets of this strange nonfeeding animal, several research groups focused their microscopes on its individual organs. There are not too many: a bright red bundle of gills, a heart with blood circulation, a muscle ring (called vestimentum) anchoring the worm in its tube. Its biggest organ is the trophosome—literally "feed sack"—and it was there that the scientists found the solution to the riddle. Each individual cell of the trophosome is populated by thousands of sulfur bacteria that, as was known already, live chemoautotrophically, meaning they can live exclusively from inorganic nutrients, building all the organic compounds and building blocks for macromolecules from scratch. These bacteria win their metabolic energy from the oxidation of hydrogen sulfide.

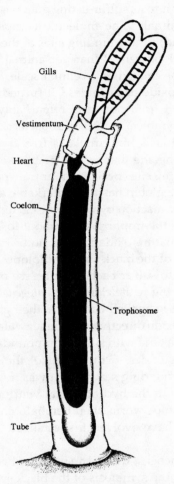

Figure 6. The tube worm, *Riftia pachyptila*, an important species in black smoker communities, does not possess any organs suitable for the intake or digestion of food, nor for excreting. It can only survive with the help of the symbiotic sulfur bacteria thriving in its trophosome.

However, the solution of this problem seemed to create a new one. The hydrogen sulfide that the tube worm obviously breathes in through its gills in order to supply it to its sulfur-eating guests is a potent poison for all previously known animals. The molecule competes with the oxygen (which all animals need) for the binding sites of the oxygen carrier hemoglobin found in red blood cells. Whatever animal breathes in hydrogen sulfide will therefore be choked on a cellular scale, in a way similar to the action of carbon monoxide or cyanides.[2] It turned out that *Riftia* hemoglobin is very different from ours and remarkably well adapted to the requirements of endosymbiosis with sulfur bacteria. The protein, which is not enclosed in blood cells but circulates freely, not only possesses an unusual oxygen-binding site that cannot be blocked by sulfide, but also has an additional binding site specialized for transporting the sulfur compound. Thus, this hemoglobin has the remarkable ability to transport the two molecules from the reaction by which the bacteria gain their energy in one ride to the cells of the trophosome without losing the high chemical energy through a premature oxidation reaction.

Other inhabitants of the black smoker biotopes also depend on symbiotic relationships with sulfur bacteria but have organized them differently. For instance, the big white clams (*Calyptogena magnifica*) populating these biotopes in scores keep their bacteria in their gills where they can get oxygen and carbon dioxide directly from the seawater. In this case, too, the way they supply the bacteria with sulfur compounds puzzled researchers. The concentration of sulfides in the blood of the clams is many times higher than in the surrounding seawater. It was found that the clams get the sulfides directly from the hydrothermal vent minerals through their foot. In contrast to the tube worm, their hemoglobin is sensitive to sulfide poisoning. The clams have evolved a specialized sulfide transporter to solve this problem.

In contrast to the lichens, which can be described as a team formed to conquer new grounds, the symbiosis of the black smoker specialists with sulfur bacteria is rather the institutionalized version of interdependences that were always present in the evolution of a small and independent biotope with short food chains. Higher organisms in the volcanic deep sea biotopes must have always depended on sulfur bacteria. In the long term, the development of "in-house" solutions may have proven economical as a shortening of supply paths by symbiosis, while the metabolisms were coupled anyway.

Endnotes

1. The term "acidic amino acid" is not in fact tautological. At neutral pH, the acidic group characteristic of all amino acids is compensated by the basic amino group, forming a so-called zwitterion with opposite charges. In proteins, these charges are neutralized by formation of the peptide bond except for the termini. Acidic amino acids possess additional acidic sites in their side chains. Only these behave like acids in proteins. There are also basic amino acids, but most of the standard 20 amino acids are neutral.

2. Luckily, the odor threshold of the notoriously smelly gas is far below the danger threshold, so that poisoning can only occur in major and sudden gas leaks.

Updates

p. 69   A paper published in 1998 assigns the heat-shock protein Hsp90 a new role as a "capacitor of evolutionary change," explaining how species can change more rapidly in times of high environmental stress. See "A Heat Shock Protein Accounts for Bouncy Evolution" in the Afterword.

p. 71   Further structural analysis of GroEL in various associations with other molecules has recently been reported. Towards the end of 1999, even a full crystal structure with a substrate mimic was published, in spite of the difficulties mentioned on page 71. At the same time, the "customers list" of GroEL was revealed. See "Who Needs a Chaperone?" in the Afterword.

4

Relevance of Extremes for Biotechnology and Medicine

This is all quite nice and interesting, you may think, but is there any use for it? Surprisingly, there is. Extremophilic microbes and the molecular mechanisms of adaptation and response to extreme conditions are not only interesting for curiosity-driven scientists who want to explore the machinery of life and its limits. Over recent years, this research has also stimulated new and exciting developments in two branches of applied science, namely biotechnology and medicine. In a way, health care is all about helping our body to cope with stress induced by illness, wounds, psychological strain, etc., so it is not surprising that medical researchers have become interested in studies of the natural stress response of the cell, which may open up new possibilities for therapy. Biotechnologists, on the other hand, are naturally curious about the vast potential hidden in the new world of extremophilic microbes. With one protein derived from an extremophilic bacterium having become a biochemical product of major importance, and another one regarded as "an extremely interesting material in search of an application," experts agree that there is a huge potential

99

for many more applications, including novel ones that would not even be conceivable without biomolecules derived from extremophiles.

Some of these promising outlooks will be presented in this chapter, but as an introduction I will first choose a historical approach to that ill-defined branch of our economy that is biotechnology.

An Extremely Short History of Biotechnology

Biotechnology is at the cutting edge of modern research and development—one can think of recombinant pharmaceutical proteins, for instance—and yet its roots date back several millennia. Ancient Egyptians knew perfectly well how to make microbes work for them, as they used yeast to make bread and beer. Recently, it has been shown that their brewers also used malted cereals, which implies that their beer was not that different from the modern product.

Although people have used microbial fermentation in brewing, baking, and winemaking for millennia, the link between fermentation and microbes was only recognized in the middle of the 19th century, when a certain Monsieur Bigo in the northern French town of Lille had serious problems with his fermentation plant producing technical-grade alcohol from sugar-beet. He called for the help of the famous chemist Louis Pasteur,[1] who had recently been appointed to a professorship at Lille. Pasteur brought his microscope and had a close look at the contents of a good batch of Bigo's production as well as some of the bad ones, which had produced a smelly mess instead of the desired alcohol. He immediately noted that the good batches had only round microbes (the yeast *Saccharomyces cerevisiae*), which in the bad batches were outnumbered by rod-shaped bacteria producing lactic acid instead of alcohol. He gave the advice to check batches for bacterial contaminations before the fermentation, and thus saved Monsieur Bigo from bankruptcy. Later, he also studied the role of microbes in the production of wine and beer and introduced a heat treatment (pasteurization) to inactivate unwanted microorganisms.

Following this landmark discovery, the science of fermentation boomed. What we now call biotechnology was known as "zymotechnology" at the time. The word was first used by the German chemist Georg Ernst Stahl (1659–1734), whose book *Zymotechnia Fundamentalis* could be regarded as the starting point of scientific investigation of fermentation.

(Sadly, the great-grandfather of biotechnology has received mostly negative publicity—he is also the author of possibly the best-known error in the history of science, the phlogiston theory.) By the end of the 19th century, zymotechnology was regarded as a key technology for the future.

One of its most remarkable success stories is linked with the then-chemist Chaim Weizmann and the foundation of the state of Israel. Chaim Weizmann (1874–1952) studied organic chemistry at Berlin and went to Manchester in 1903, where he first worked on synthetic dyes and then on fermentation procedures. During World War I, he was asked to find an inexpensive way to produce the solvent acetone, which was urgently needed. He solved the problem within weeks by identifying a bacterium, *Clostridium acetobutylicum*, which can degrade cereal starch to acetone and butanol in a single fermentation step. (Earlier methods required separate steps for the hydrolysis of starch to glucose and for the fermentation of glucose.) In recognition of his contribution to British biotechnology and the war effort, he was allowed to participate in the preparation of the Balfour Declaration (1917), which eventually led to the foundation of the state of Israel with Weizmann as the first president (1948–1952).

Only a few years later, another fermentation process was introduced that is still in use today, namely the microbial production of citric acid, which is required by the food industry in ever-growing amounts. The first commercial production of the acid from lemon juice was started by the brothers John and Edmund Sturge in Selby, England, in 1826. In 1917, however, it was found that the mold *Aspergillus niger* can be used to produce citric acid. Large-scale production began in 1923 in Brooklyn, New York.

Today, there is a wide range of products made either with enzymes or with living microorganisms on an industrial scale. There are small molecules—like ethanol, produced as a fuel in Brazil to reduce the country's dependence on oil imports—but also peptides and proteins, which become more and more important for pharmacology and cannot be produced chemically at reasonable cost. The development of molecular cloning techniques in the 1980s has very quickly moved up to the technological scale, leading to a boom with dozens of small biotech firms newly founded in America and elsewhere. Thus, cloning has become one of the keywords that today most people associate with biotechnology.

But why should biotechnologists be interested in extreme conditions? For various reasons, including both technical and microbiological considerations. First, microbiologists tend to be haunted by the fear of contaminants. In particular, if they grow microorganisms under "normal" conditions (20 to 37°C, neutral pH, rich medium) they can be sure that any germ dropping in from the air will grow as well as (or even better than) the species of interest. Doing microbiology on a technical scale, they have to exclude contamination with 100 percent reliability, which can complicate procedures enormously. However, if one could conduct the desired fermentation process at extreme pH or temperatures, say, near the boiling point, the risk of contamination would be reduced considerably. Second, from the technical point of view, chemical engineers like to vary parameters such as temperature, pH, or pressure to optimize the yield of their processes. For instance, a given process involving both enzyme-catalyzed reactions and inorganic catalysts might be most economical when conducted at 100°C and 200 bar. Since the discovery of hyperthermophiles, such conditions have become a realistic option in biotechnology.

Hyperthermophilic Enzymes

Enzymes are tremendously useful—not only for cells. Stain removers, "biological" washing powders, and some cosmetics contain enzymes, and various kinds of food are produced with their help. Many other processes from daily life to industrial-scale production rely on enzymes. And for any given process enhanced by the action of an enzyme, one could, in principle, find a heat-resistant variant of the biocatalyst that would be able to do the same thing at high temperatures.

Application of such "extremozymes" has only just begun. Their potential has been recognized, and perhaps a dozen of them have been successfully expressed (produced using cloning methods) in mesophilic (i.e., "normal") bacteria, so that they can be produced and purified on a technical scale. The extremozyme most widely used today is the thermophilic DNA polymerase used for the polymerase chain reaction in molecular biology labs all over the world (see the Profile on Kary Mullis).

Up to now, none of the extremozymes is being used in industrial production, but there is no shortage of candidates, interest, and new ideas. For instance, the isomerization of glucose leads to an equilibrium mixture of the isomers glucose and fructose, with the fructose content increasing with temperature. It has been shown that, using the thermostable glucose isomerase from *Thermotoga maritima*, one can conduct the process at 100°C

 profile _____

Kary Mullis and the Polymerase Chain Reaction

A bright idea and a heat-stable enzyme—these were the only two ingredients needed to develop the polymerase chain reaction. The idea fell upon the chemist Kary B. Mullis during a long car ride one spring night in 1983. It is quite simple: If you copy a piece of DNA (using an enzyme called DNA polymerase), heat the newly built double helix so the two strands fall apart, cool it, copy it again, heat it, and so on, you should be able to double the number of copies in each cycle, starting from as little material as, for instance, one individual molecule. Without even knowing the sequence of this molecule, you could increase your amount of material so much that you could do anything with it—for instance, determine the sequence, transfer it into bacteria to make them produce the corresponding protein, or put it in bottles and sell it.

Mullis, then a nucleic acid chemist with the small company Cetus, saw this potential, and he couldn't believe his luck. He haunted colleagues with his idea trying to find a rub, or an old report saying someone tried it and it didn't work. As a matter of fact, no one had ever tried or even thought of it, and, more amazingly, several of his colleagues initially failed to see the point.

So much for the idea. We now need an enzyme (a DNA polymerase, to be specific) that is stable at 90°C (the temperature needed to disintegrate the double helix) and optimally active at not too low a temperature, so that the temperature cycling can be carried out relatively quickly. No problem at all. Every living organism has a DNA polymerase, and hyperthermophilic archaebacteria, of course, have extremely thermostable DNA polymerases. As the structure of DNA is exactly identical in all living beings (at least on this planet), hyperthermophilic DNA polymerases can easily copy DNA from any organism. The first enzyme that was widely used for the polymerase chain reaction (now universally known as PCR) was from *Thermus aquaticus*, one of the first hyperthermophiles that Brock found at Yellowstone. It was patented for Cetus in 1989 and became commercially available as Taq-polymerase.

Mullis' discovery started a genuine revolution in molecular biology. In the times before PCR, the genetic text encoded in DNA could only be read if an enormous number of identical molecules was at hand. A very small sample by these standards might contain one picomole (10^{-12} mol), corresponding to more than 10^{11} molecules. However, since PCR has become available, any sample will do. One only needs to know (or guess) a very small part of the sequence (which may actually be in the region next to the unknown gene of interest) in order to design a set of starter molecules. As long as one is dealing with a well-known species, this is no serious restriction. For instance, one could remove invisibly small tissue samples from ancient mummies and design the starter oligonucleotides on the basis of genetic data of today's human beings.

PCR can also help the police. Minute blood splotches, worn-off skin flakes, or other microscopic tissue samples are sufficient to determine the genetic identity of a person. Furthermore, the Human Genome Project, which is set to decode the entire sequence of all our genes, would have been inconceivable without PCR. All these things (and a couple of others, short of the cloning of dinosaurs *à la Jurassic Park*) have been made possible by the use of a thermostable enzyme.

Kary Mullis received the Nobel prize in chemistry only 10 years after his late-night inspiration. He gave up research and achieved notoriety for his eccentric behavior at international conferences. He recently started a company to amplify the DNA of dead stars like Elvis Presley or Marilyn Monroe (not to revive them, just for sale as molecular souvenirs).

(rather than at 60°C) and thus achieve a considerably higher yield of the desired product, fructose.

A more unusual application of thermostable enzymes has been suggested by Jonathan Woodward and his colleagues at the Oak Ridge National Laboratory in Oak Ridge, Tennessee: the generation of energy from renewable sources. The combined application of the enzymes glucose dehydrogenase from *Thermoplasma acidophilum* and hydrogenase from *Pyrococcus furiosus* would allow the production of gluconic acid and molecular hydrogen from glucose, which could be obtained from waste paper, starch, or other renewable sources. As yet, there is no detailed analysis of the commercial practicality of the process. At present, gluconic acid is more expensive than glucose, suggesting the authors have conceived a bifunctional cash-cow, which would make profits from both products of the reaction. But this will only work out if both enzymes and the required coenzyme NADPH (the reduced, hence H for hydrogen, form of nicotinamide adenine dinucleotide phosphate) are kept stable for a sufficiently large (in fact, astronomically so!) number of turnovers. Presumably, the high cost and poor stability of the coenzyme will be the limiting factors.

Further potential applications of extremozymes are anticipated in fields such as the chlorine-free bleaching of paper and pulp products, oil production, and food processing.

Preservation by Freezing and Freeze-Drying

When the mass production of electrical refrigerators began in the 1920s, cooling rapidly became the most common and simplest method to preserve food in the household. It owes its efficiency to a combination of effects. Chemical reactions tend to occur more slowly in the cold, and the activity of food microbes is also slowed upon cooling and may be stopped altogether if the food is frozen and kept at sufficiently low temperature. On the other hand, cellular structures of the food itself may suffer from freezing and thawing. This leads to fruit and vegetables losing their texture and rigidity (as you will know if you ever had an overachieving refrigerator freezing cucumbers and strawberries). It needs a good deal of experience to make decisions on which food can be frozen and which cannot.

More generally, if we want to have food with a reasonably long shelf-life, we have to have it processed with some kind of extreme condition to kill off microbes. We can have our peas boiled for an hour and then canned, or we can have them frozen. The question is which method is best suited to avoid damage to the cells of the food itself, to its color and flavor, and to its vitamins. I will explain this in more detail in the following section, which deals with application of pressure in food processing.

In a similar way to food, drugs are another important group of products that have to be protected from chemical and microbial decay. Ironically, this has become more difficult due to the fast progress that pharmacology has made over the past few decades. Whereas classical pharmaceuticals are small molecules, typically enzyme inhibitors like aspirin, there is now a trend in pharmacology toward macromolecules (mainly proteins, such as insulin, clotting factors, interleukins). However, the new, "more biological" drugs are more perishable than the classical small-molecule prescriptions. Freeze-drying used to be the method of choice for their preservation. In this process, an aqueous solution of the substance is rapidly frozen (e.g., by immersion into liquid nitrogen) and then subjected to high vacuum. The solvent is thus evaporated directly from the solid state (sublimation). What remains is a very porous, light powder, the best-known example being instant coffee. Many proteins commonly used as laboratory chemicals are shipped as freeze-dried powders, and with the most stable proteins one can repeat the freeze-drying and redissolution indefinitely without any harm.

However, recent investigations carried out in the cryopreservation division of the company Pafra (whose director, Felix Franks, is a world authority in the field) suggest that the cold treatment may not be gentle enough for some sensitive preparations. It is true that all reactions are stopped once the sample has uniformly acquired the temperature of liquid nitrogen, but during the cooling, solutes can enrich in the environment of the biomolecules, because initially the solid phase will only contain pure water. If the local concentrations of dissolved chemicals are increased by several orders of magnitude, this can enhance reactions that would normally be very slow and enable them to cause damage even at low temperatures and in the short time needed for shock freezing.

Therefore, Franks and his co-workers look at alternatives to freeze-drying. Their current favorite is the glass state, a supracooled liquid that (like common window glass) behaves like a solid although it does not have an ordered structure. The extremely high viscosity in glassy materials means that molecules can only move incredibly slowly, and hence only react slowly. To get an idea of the time scale, consider that ancient window panes in medieval churches are slightly thicker at the bottom, as it took the glass a couple of centuries to flow downward. Similarly, the formation of the proper solid state, namely crystals, takes time. Glasses with little sparks of crystals are typically found in museums of Roman history, indicating that these states are sufficiently stable for the duration of the shelf-life desired for drugs.

profile _____

Pierre Douzou and the Invention of Cryoenzymology

"Vous cherchez quoi au juste?" (What exactly are you looking for?)—
this seems to have been a frequent response when Pierre Douzou
specified his profession, for a researcher in French is a *"chercheur."*
Vous cherchez quoi au juste is also the title of his autobiography published in 1994. Most of the time, the son of a glovemaker from the
Provence knew what he was looking for, even if his career leading him
from the sanitary services of the French army to the national science
organization Centre National de la Recherche Scientifique (CNRS)
and eventually to the Museum National d'Histoire Naturelle wasn't
always a straight line.

He searched for and found a certain method that eventually
made him the mentor of French research on extremophiles and extreme conditions, namely cryoenzymology. Toward the end of the
1960s, when classical enzymology was about to be brushed aside by
the rapidly advancing new molecular biology, Douzou suggested to
try to grasp the fleeting reaction intermediates of fast enzyme-
catalyzed reactions by cooling. According to an empirical law stated
by the Swedish chemist and winner of the third ever Nobel prize for
chemistry, Svante Arrhenius (1859–1927), chemical reactions run
faster at higher temperatures and slower at lower ones whether they
are catalyzed or not. But the range between the normal room temperature of 20°C and the freezing point of water is not sufficient to
slow reactions considerably nor to stabilize labile intermediates on a
longer timescale. In order to obtain subzero temperatures, Douzou
had to add antifreeze agents to the solutions. However, when he
tested several alcohols that would have been suitable from the physical point of view, he found that they all denatured the enzymes, and in
some cases the damage was irreversible.

At this point, Douzou came across a paper in a journal of ento-
mology (the science of insects) showing how insect larvae can avoid

freezing at temperatures as low as $-60°C$. Their cells produce high concentrations of glycerol, up to 50 percent by volume. At ambient temperature, this concentration would be fatal, but the larvae seem to have evolved a strategy of gradually matching the glycerol concentration to the sinking temperature to avoid damage. Inspired by their survival strategy, Douzou and his co-workers developed a methodology that they presented in 1970 under the name of cryoenzymology.

Like the insect larvae, cryoenzymologists balance the influence of cold and medium composition in such a way as to keep the enzymes intact and active at temperatures way below the freezing point. Due to the temperature-dependent reduction of reaction rates described by Arrhenius, researchers can cut down complex consecutive reactions into individual steps and measure their rates individually. It took Douzou and his group two years to collect the fundamental physicochemical data concerning their antifreeze mixes and to prove that they do not denature the enzymes. Only then could they start to acquire real data on enzyme kinetics. The first enzyme they chose to study was horseradish peroxidase.

In the following two decades, the methodology was picked up by several laboratories and applied to dozens of enzymes. Often, the antifreeze has to be specifically adapted to the nature of the enzyme to be studied. The most commonly used additives are glycerol, methanol, and dimethyl sulfoxide. Even structural studies by X-ray crystallography as well as by NMR have been carried out at subzero temperatures. Generally speaking, the combination of enzymology with extreme conditions (including high pressures and temperatures) has been particularly fruitful in the hands of the present generation of French *chercheurs*, many of whom have worked with Douzou at some stage.

But how do we get our pharmaceutical preparation to form such a glass state? Basically by removing the solvent and keeping the material from crystallizing. In most cases, this can be achieved by adding carbohydrates (like sugars, or starch) to the solution, which will then form a kind of syrup upon evaporation of the solvent. Although such a method has not yet been formally approved for pharmaceuticals, it is well established in nature—by insects that use this approach in making their larvae frost-hardy (see the Profile on Pierre Douzou). In the next section, I will continue the preservation-sterilization theme and proceed from freezing to squeezing.

High-Pressure Biotechnology

In the beginning there was the egg. Percy W. Bridgman (1882–1961), an American physicist who developed instruments for the use of high pressures in the laboratory (including the Bridgman cell) and was awarded a Nobel prize in 1946, treated an ordinary egg with a pressure of 7,000 atmospheres for 30 minutes. After that, the egg was as hard as if it had been boiled. This experiment, which was first published in 1914, demonstrates two things: First, that proteins can be denatured by pressure, as denaturation and aggregation of egg proteins are the molecular reasons underlying the hardening of a boiled (or pressurized) egg, and second, that heat treatment of food can be replaced by pressure treatment.[2]

In fact, pressure can not only replace, but even outperform heat, as an exposure to high pressures does not accelerate unwanted side reactions as high temperature tends to do. With eggs, for instance, overcooking can lead to a sulfury taste, while no length of pressure treatment would do this. This is because virtually all chemical reactions can be accelerated by an increase in temperature in a mathematically uniform way first described by Arrhenius (see the Profile on Pierre Douzou). In food processing, chemical reactions triggered by heating may well destroy molecules that are crucial for the flavor or color of the food. In contrast, pressure only affects reactions that are coupled with a significant change in the overall volume of the system, and thus rarely induces the making or breaking of stable (covalent) chemical bonds. Typically, pressurization will reversibly separate loosely associated molecular assemblies, for instance, complexes consisting of several protein subunits. It will also unfold the three-dimensional folded structure of proteins, which is actually a process desired in cooking, as in the egg example.

Apart from the denaturation of proteins, food technologists can also make use of high pressures for the inactivation of microorganisms (preservation of food), gelatination of starch (for jam production), and the inactivation of enzymes, which might catalyze unwanted side reactions. In

Japan, high pressure food processing has been used for years—most successfully in jam fabrication, where both the gelatination and the sterilization effects of pressure are useful. Since 1990, several brands of purely pressure-processed (that is, never heated) jam have been available in supermarkets. The pressurized jams are much more like the untreated fruit than the conventionally cooked products and have therefore found a steadily growing circle of friends, despite their higher prices.

In Britain, the recent series of food safety scares ranging from "mad cow disease" to meat contaminated with *Escherichia coli* has raised fresh interest in pressure methods as an alternative in food sterilization. Killing ordinary bacteria with pressure is relatively straightforward, and its future will depend mainly on economic considerations—whether heat or pressure sterilization equipment will be cheaper to buy and maintain. The questions get trickier when it comes to spore-forming bacteria, which, as I mentioned in the previous chapter, require higher sterilization temperatures. Spores are also remarkably resistant against a single pressure treatment, but they can be tricked into giving up their protected state and starting to germinate by a suitably designed series of pressurization steps. Once the spores have germinated, a simple pressure shock of a few thousand atmospheres will kill them off.

In Japan, where the notion of technologically processed food does not scare customers as much as it would in Europe, further fields of potential application are being developed, including, for instance, fresh vegetables, fruit juice, meat, milk and milk products, coffee, and tea (Table 1). The research and development in high pressure food science pioneered by Rikimaru Hayashi of Kyoto University since 1986 is still more active in Japan than anywhere else.

One of the products that may be improved in the future by processing methods developed in Hayashi's laboratory is milk for infants. There is a major difference between bovine and human milk, which would become obvious if one tried to make cheese of both. When one adds the rennet, one kind of milk protein called casein precipitates and ends up in the cheese. A mixture of other kinds of protein remains in the liquid phase, the whey. Human milk contains roughly equal amounts of casein and whey protein, while cow's milk contains four times more casein. While this is good for cheesemaking, it is not necessarily good for young babies. Thus it appears logical to use whole milk and add some extra whey to produce milk for

Table 1. Research and Development in the Field of High Pressure
Use in Food Technology in Japan, 1987–1992[a]

Application	Objects investigated
Food sterilization	Meat, fish; egg white; milk and milk products; tomato juice; jams; tea drinks; fresh vegetables; tangerine or satsuma juices; plums; fruits; pickles; coffee; sausage
Processing of meat and meat products	Ultrastructure and myofibrillar proteins of beef muscle; properties of myosin B; gel formation of myosin; properties of pressurized meat
Processing of fish products	Gel formation of Alaska pollack; properties of pressurized fish meat; surimi; water-soluble fish meat; sea urchin
Fruit and vegetables	Development of jams; fruit sauces and desserts; extraction of pectin; processing of plums; enzyme inactivation in oranges; processing of orange juices; control of bitter taste in grapefruit juice; properties of pickles; properties of soy protein
Processing of other foods	Texture of pressurized egg white; cheese processing
Cooking	Egg; meat; fish; oyster; shrimp; Japanese radish

[a]From Hayashi, R. "Utilization of pressure in addition to temperature in food processing and technology," in *High Pressure and Biotechnology*, C. Balny, R. Hayashi, K. Heremans, and P. Masson, eds. (London: John Libbey, 1992), pp. 185–193.

infants. Alas, bovine whey contains a high proportion of a specific protein, beta-lactoglobulin, that may cause milk allergy in infants and should be removed. While several methods tested for this purpose have proven too unspecific, Hayashi and co-workers have shown that the protein-digesting enzyme thermolysin, if applied under high pressure, specifically chops down the offending globulin and leaves the other whey proteins intact.

Beyond food technology, there is further potential that is not yet being realized. Gases, e.g., carbon dioxide, pressurized beyond their critical point (the temperature and pressure where the difference between gas and liquid state disappears) can be very efficient solvents for the extraction of natural products. Furthermore, the well-studied effects of pressure on cells and individual enzymes could be used in biotechnological processes. In these areas, there is a lot left to explore.

Bacteriorhodopsin as an Optoelectronic Component

Being able to thrive in saturated salt brine is by no means the only peculiar trait of the halobacteria we met in Chapter 2. Their cell membrane (the layer of lipids and proteins surrounding the cell) is extraordinary as well. Walther Stoeckenius couldn't know just how peculiar it was when, in 1964, he and his group set out to investigate it. In their first attempts to fractionate the membrane and purify individual components, they were struck by the properties of an intensely purple fraction that Stoeckenius called the purple membrane.

The purple membrane, which occurs in patches on the surface of live bacteria, turned out to be extremely rich in protein, with only one-quarter of its weight being lipid. Moreover, the protein component occurred in an ordered array that can only be described as a two-dimensional crystal. Dieter Oesterheld joined Stoeckenius, who by then had moved to the University of California at San Francisco, to study the molecular composition of this natural crystal. He found that it contains only one protein, which at 26,000-dalton molecular weight is not even particularly large. This protein, named bacteriorhodopsin, has kept Oesterheld busy for more than two decades. Small surprise, as it turned out to be one of the most fascinating biological materials that scientists have ever discovered, and the most promising candidate for the development of biomolecular computers.

Although bacteriorhodopsin is distantly related to the visual pigments of animals, the rhodopsins (and also contains the pigment molecule retinal), it serves an entirely different purpose. Halobacteria use it to acquire energy from sunlight, especially when oxygen becomes scarce. When bacteriorhodopsin catches a photon with its retinal "antenna," this induces an isomerization of retinal (Figure 1), which then becomes detached from the protein and thereby triggers conformational changes in the protein itself. These eventually lead to the complex changing color and pumping a proton across the membrane. This way, bacteriorhodopsin generates an electrochemical potential across the membrane, which is simply a way to store energy. What makes bacteriorhodopsin interesting for potential applications in bioelectronics is not only the vast number of intermediates on its reaction cycle (Figure 2), each of which can be selectively addressed by light of a specific wavelength, but also its extreme stability (up to 140°C) and durability. Each molecule of bacteriorhodopsin can run through the photocycle more than 10 million times—an achievement unmatched by any synthetic material.

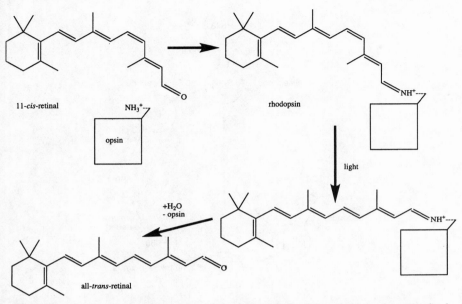

11-*cis*-retinal

opsin

rhodopsin

light

+H₂O
- opsin

all-*trans*-retinal

Figure 1. The isomerization of the kinked 11-*cis*-retinal (top left) to the stretched version all-*trans*-retinal is crucial for both the visual process in our eyes and the energy uptake in the photosynthetic pigment bacteriorhodopsin.

When these remarkable properties of bacteriorhodopsin became known at the beginning of the 1970s, Soviet scientists were the first to consider bioelectronic applications. Through these, they hoped to compensate for the Western advantage in semiconductor electronics. Parts of the research and development performed in the Soviet Union are still classified. It is known, however, that a microfiche material containing bacteriorhodopsin was developed.

However, successful applications did not turn up as quickly as one would have imagined from knowing the promising characteristics of the substance. Countless suggestions have been made, many ideas have been developed into experimental processes, only to fail commercially. Ironically, the most spectacular defeat was encountered in the quest for an application matching the natural task of the protein—converting sunlight to chemical or electrical energy. If one was to use purple membrane extracts to generate solar electricity, the energy yield would be just 1 percent, compared with 14 percent achieved by modern photovoltaic cells. Still, a

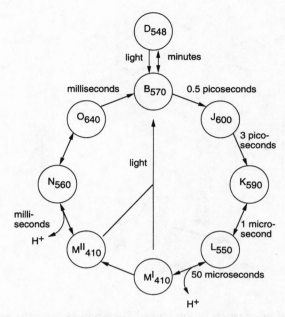

Figure 2. The photocycle of bacteriorhodopsin contains the most important states of the molecule, which can be characterized by spectroscopic methods. Each state is assigned a capital letter, with the subscript number indicating the wavelength of maximal light absorbance. The availability of conversions that can be triggered by different wavelengths and occur with different rates makes bacteriorhodopsin a variable and promising material for future optoelectronic elements.

niche application might be found in the regeneration of the biological energy carrier ATP from ADP (adenosine diphosphate). Biotechnological processes using a lot of the expensive ATP might benefit from a solar-powered recycling system.

Prospects are much better in those fields where bacteriorhodopsin is used as a photochromic material (i.e., a substance that reversibly changes color after illumination). Possibilities range from optical data storage to holographic memory with associative functions. Computers equipped with such a memory could, for instance, recognize a face or a fingerprint in one step, without time-consuming one-by-one comparisons with stored pictures. Robert Birge, who is developing such systems with his group at the Keck Center for Molecular Electronics in New York, predicted in 1995 that hybrid computers containing both bioelectronic and semiconductor elements "will be available within eight years." Those, says Birge, would be the first computers to fulfill all requirements for artificial intelligence.

As yet, bacteriorhodopsin and hyperthermophilic DNA polymerases are the only enzymes from extremophiles whose application has received widespread attention. From the same sources, dozens of other, equally fascinating and promising biomolecules could be obtained. Watch this space . . .

Extremophiles and Disease:
Acid-Resistant Bacteria in the Stomach

Extreme conditions like high pressure, high temperature or acidic medium can be used to kill microorganisms for the purpose of food preservation, as we saw in the first sections of this chapter. To a limited extent, our body can use similar methods. Thus, one of the functions of the acids in the saliva and stomach is to destroy bacteria taken in with the food. However, just as halobacteria can thrive in salt meat and *Deinococcus radiodurans* can survive gamma-ray sterilization, there are other bacteria that can make themselves comfortable in the hostile environments of the mouth (like those that keep dentists employed) and even in the stomach.

Much like the extremophile hunters in their field studies, the Australian pathologist J. Robin Warren found bacteria in a place where they should not be able to live according to textbook wisdom, namely in the human stomach. The spiral-shaped bacteria later classified as *Helicobacter pylori* had retreated into the mucus layer that covers the walls of the stomach. Only after a series of failures did Warren and his co-worker Barry J. Marshall manage to cultivate the new kind of bacteria. When they published their results in 1983, researchers around the world confirmed the occurrence of such bacteria in the stomach, especially in patients with chronic superficial gastritis, a condition involving a persistent inflammation of the stomach.

However, the presence of bacteria in a diseased tissue does not prove them guilty of causing the disease—they might just have profited from the body's weakness and invaded an organ already afflicted by disease. Marshall and another volunteer did a self-test to find out whether the presence of the bacteria was the cause or a consequence of the disease. The two healthy men swallowed a dose of *Helicobacter pylori* and indeed both became ill with gastritis. Obviously, the infection with *Helicobacter* nearly always leads to a superficial gastritis, which, however, may often be overlooked and blamed on a heavy meal, for instance. If the infection persists and is not treated in time, it can lead to ulcers of the stomach or the duodenum in the long term (Figure 3).

This finding overturned a dogma that was almost as old as Western civilization, namely that ulcers are caused by excessive acid production by the stomach. In the first century AD, the Roman physician Celsus recommended low-acid food against ulcers. Since the 1970s, there have been drugs that reduce the acid production of the stomach without major side effects, and indeed reduce ulcers. However, when the treatment is stopped,

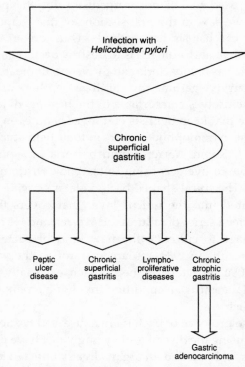

Figure 3. Potential consequences of an infection with *Helicobacter pylori*. The sizes of the arrows symbolize the different probabilities of the subsequent illnesses.

the ulcer always comes back. In contrast, a treatment with bismuth prescriptions or antibiotics that eradicates the *Helicobacter* population can heal the gastritis permanently.

But how do the bacteria manage to settle in the human digestive tract without getting digested? The secret seems to lie in their mobility and in some chemical specialties of their metabolism. Mobility is crucial when the contents of the stomach get flushed down to the guts. With the help of their flagella, the spiral-shaped bacteria can swim fast enough to escape the fate of ending up in the loo. And the special "trick" of their metabolism is that they produce enormous amounts of the enzyme urease, which can degrade urea (a product of the digestion of proteins) to form ammonia and carbon dioxide. One possible explanation of the acid resistance of *Helicobacter* is that the bacteria may be able to use the ammonia produced by urease for the neutralization of the gastric acid in their immediate environment.

This case of adaptation to extreme conditions is immensely important for global health considerations. Scientists have estimated that one-third of the world population carries a latent infection with *Helicobacter*, but as with the tuberculosis germ, only a part of the infection cases lead to recognizable illness symptoms. About 10 percent of all human beings develop ulcers at some stage of their life. Both for ulcers and for stomach cancer, a clear correlation with the number of individuals infected with *Helicobacter* can be found in comparative studies. Infections and the diseases now believed to be their long-term consequences are more common in developing countries than in industrialized ones, and they have both been declining slowly over the course of the 20th century. A large-scale campaign to fight the germ could prove a very efficient measure against ulcers and cancers of the stomach and the duodenum.

Medical Applications of Heat Shock Proteins

Although stress-resistant bacteria can cause a lot of trouble, there is also a realistic chance of using parts of the cellular stress response to cure disease. The growing interest of biochemists in those heat shock proteins that act as molecular chaperones in the cell has also infected medical researchers. According to their preliminary results, heat shock proteins may play important roles in a variety of pharmacological applications in a not too distant future.

The heat shock protein Hsp70, for instance, turned out to be the target molecule of the immunosuppressive drug desoxyspergualin (DSG), which is currently being tested for use after transplantations. Preliminary results from the laboratories of the drug company Bristol-Myers Squibb suggest that DSG binds to the heat shock protein in a way that inhibits its normal interaction with another heat shock protein, Hsp40.

Another heat shock researcher, Pramod Srivastava of Fordham University, New York, believes he can use heat shock proteins to cure cancer. His method involves isolating the heat shock protein Hsp96 from tumor cells and reinjecting the purified protein into the patient's blood. Results from animal experiments and preliminary clinical trials suggest that this approach not only can cure the acute cancer, but can also immunize the patient against further attacks of the same kind of tumor. Obviously it is not the heat shock protein as such that produces this remarkable effect—in this case the purification and reinjection procedure would make little sense. The clue seems to lie in Hsp96's unique ability to bind tumor-specific peptides while in the cancer tissue, keep hold of them during purification, and present them to the immune system upon reinjection. Although heat shock proteins from healthy human tissue do not stimulate the immune system, the cancer-hsp with the bound peptide seems to act as a strong adjuvant (an agent stimulating the immune response against something else) in that it induces an immune response that is many times stronger than the one triggered by the peptide alone. This may not be such a big surprise after all, as one of the most commonly used and most efficient stimulators of the immune system, Freund's adjuvant, contains heat shock proteins from mycobacteria.

There are other ways to use the immune-stimulating potential of heat shock proteins. In 1994, the company StressGen in Victoria, Canada, which has been producing chemicals for heat shock research since 1990, opened a research department for the development of vaccines and adjuvants based on combinations of heat shock proteins with antigens (i.e., molecules or parts of molecules which trigger an immune response). Researchers at StressGen follow three main lines:

■   Genetic coupling of heat shock proteins with known peptide or protein antigens to form fusion proteins

- Chemical cross-linking of antigens with heat shock proteins
- The exploitation of the natural tendency of heat shock proteins to bind certain (partially) unfolded proteins and peptides

In all three research areas there is now some hope that heat shock proteins can be used as efficient and unproblematic enhancers of the immune response. There is definitely a demand for such formulations—in 1996, there was only one medical adjuvant on the U.S. market.

A second, equally important field for potential applications of heat shock proteins lies in the prevention of tissue damage as it occurs after ischemia (the lack of supply with arterial blood after a stroke or heart attack). In a tissue that has been deprived for some time of oxygen-containing blood, the reestablishment of the oxygen supply can trigger a reaction cascade that can cause severe tissue damage. This effect is blamed for the major part of the brain damage in stroke patients.

As early as 1986 some researchers suggested that heat shock proteins might help to prevent this damage. In 1995, Wolfgang Dillmann at the University of California at San Diego obtained the first real evidence for this hunch, when his group showed that mice genetically manipulated to overproduce Hsp70 suffered 40 to 50 percent less damage after experimentally induced ischemia. Should this protective effect be found in humans as well, researchers would have to find a method to supply patients with a high risk of stroke or heart attack with a protective dose of heat shock proteins. While cell biologists normally induce them in cell cultures by applying some kind of stress or by switching on a manipulated gene, some kind of stress-free and noninvasive supply would have to be found for medical applications.

One researcher who has already looked for drugs that can induce the heat shock response without unpleasant side effects is Richard Morimoto of Northwestern University in Evanston, Illinois. He believes that existing anti-inflammatory drugs are just right for this purpose. One of these, indomethacin, apparently can help to prevent cell damage by acting on one of the most important regulators of the heat shock response, the heat shock factor Hsf1.

In the future, the folding helper function of those heat shock proteins known as molecular chaperones may also become medically relevant. More and more diseases that at first appeared mysterious are now blamed on molecular processes that can be ultimately traced back to misfolded proteins. For instance, the plaques held responsible for the brain damage suffered by patients with Alzheimer's disease are known to consist mainly of aggregated protein. One of these, the human variant of the egg protein lysozyme, can be turned from a normal protein into an aggregating, plaque-forming one by the mutational exchange of a single amino acid. Similarly, the prions causing the "mad cow disease" (more formally known as BSE, for bovine spongiform encephalopathy) are believed to consist exclusively of protein. Even though it is not yet quite clear how an infectious prion protein manages to convert a healthy prion precursor protein, many researchers believe that alternative folding modes play an important role in this process. Considering the importance and urgency of the question of whether there is a connection between BSE and the human variant, Creutzfeldt–Jakob disease, elucidation of the mechanisms of action and infection of prions is a research goal of top priority.

To what extent and in which ways misfolded proteins can cause disease are among the hottest questions of the current research on the borderline between biochemistry and medicine. Should scientists succeed in tracking down the offenders, molecular chaperones may find a new role in helping doctors to restore law and order in the convalescing human body.

Endnotes

1. Louis Pasteur (1822–1895) discovered the chirality ("handedness") of molecules as well as the importance of microorganisms for fermentation processes and infectious diseases. He developed sterilization methods (pasteurization) and vaccinations against various diseases and founded the Institut Pasteur in Paris. His overwhelming and enduring influence has also been criticized, most notably by Bruno Latour, who fears the "pasteurization of French science."
2. The very first recorded example of pressure-treated food dates back to 1850. An employee of the Jardin du Luxembourg named Masson compressed dried vegetables using a hydraulic press. However, he was not intending to study the effects of pressure, but rather to save packing space onboard ships.

5

Extremists and the Tree of Life

In the Introduction I defined—as far as possible—what we understand by "normal" conditions. But today's normality has not always been so. It is true that the climate on our planet has been amazingly constant for the past three and a half billion years—a fact that is cherished as evidence for self-regulation by the supporters of James Lovelock's Gaia theory, as we will see at the end of this chapter. However, when life on Earth began, the planet certainly was what we would consider a hostile environment by today's standards. Thus, it is quite possible that some of the organisms that we know as extremophiles have done nothing special beyond sticking to their three-billion-year-old habits. Hence, investigations of their special habits may offer a unique window into life's distant past.

Therefore I will now go back into the "dark ages" of life's history, into the time before evolution came up with plants and animals. From the origin of life to the emergence of the first multicellular organisms, there are many intriguing findings and even more open questions to report. And extremophiles play an important part in this.

The Origin of Life—
the Primeval Earth as an Extreme Habitat

Our planet as we know it—with the layer structure including a liquid core, a semisolid mantle, and a crust including continents and the sea floors—is some 4.4 billion years old. It arose from the aggregation of cosmic dust and meteorite impacts. Soon after the internal structure of the newborn planet had stabilized, gas emissions from the core started to form the primeval atmosphere, possibly within as little as one million years. Its composition was nowhere near the gas mix we now call air. Nitrogen was the main component as it is now, but it was followed in abundance by water vapor, carbon dioxide, methane, and ammonia. In stark contrast to our current oxygen-rich atmosphere, which is oxidizing (so it sustains fire and makes iron rust), the primeval atmosphere was reducing and virtually devoid of oxygen (Figure 1). Without oxygen it

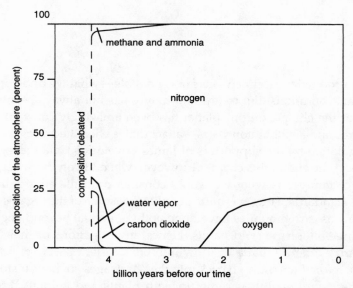

Figure 1. Changes in the composition of the Earth's atmosphere since its beginnings. At the present day, dry, clean air contains the following main components (volume percentages): nitrogen, 78.01; oxygen, 20.95; noble gases, 0.93; carbon dioxide, 0.03.

could not form an ozone layer and thus provided little protection from the hard UV radiation of the Sun.

Some time after the internal structure of the planet was in place, continents started to form. The oldest rocks dated with isotope methods are zircons from Western Australia formed some 4.1 to 4.2 billion years ago, a value accepted as the minimum age for the continents. At this time, the young planet was frequently battered by huge meteorites, which could be up to 100 kilometers in diameter. The energy of such an impact must have heated most of the seawater to boiling and saturated the atmosphere with dust for millennia—surely not the ideal kind of condition to start a biosphere. The meteorite showers faded some 3.9 billion years before our time. By geologists' standards, the oldest fossils of cellular life forms are only a little younger than that. They prove that microorganisms resembling today's cyanobacteria lived as early as 3.5 billion years ago. Gustaf Arrhenius and his co-workers at the Scripps Institute for Oceanography in La Jolla, California, even believe they can push the limit back by 400 million years. Using a new kind of mass spectrometry to determine the isotope ratio in our planet's oldest known sediments, a 3.8 billion-year-old formation from West Greenland, they found evidence for biological activity that took place in these materials before they were cooked and pressurized to stone.

The somewhat surprising conclusion to be drawn from these studies is that as soon as the Earth was a reasonably stable body with oceans, continents, and an atmosphere, and the meteorite impacts became less frequent and less catastrophic, the evolution of life began. At that time, the luminosity of the Sun presumably was only three-quarters of what it is today, but the greenhouse effect of the carbon dioxide-rich atmosphere must have overcompensated for this difference. However, the details of the climate changes during the first two billion years of our planet remain controversial. Inorganic mechanisms may have started reducing the carbon dioxide content of the atmosphere (while the Sun's luminosity increased) even before evolution came up with photosynthesis. As the oxygen-free primeval atmosphere could not provide much shelter from the fatal UV radiation of the Sun, life could only originate underneath a protective layer of water or soil.

So much for setting the scene and the basic facts. What really happened between the consolidation of the Earth's crust (4.2 billion years ago) and the appearance of cyanobacteria (3.5 billion years ago), and how a planet bubbling over with volcanoes and battered by meteorites became an oasis of life within a largely lifeless universe, remain open to speculation. Many steps on this pathway require an explanation. All biological material is built from relatively small and simple building blocks—organic molecules (such as amino acids) that can be assembled to form long chains, known as biological macromolecules or biopolymers (such as proteins). However, most of these building blocks would not be there if there weren't any organisms on Earth to make them. Thus, we are facing a classic chicken-and-egg problem, but luckily the famous Urey–Miller experiment provided some clues as to where the first building blocks came from (see the Profile on Stanley Miller). In the next step, the building blocks must have assembled to form macromolecules, and those must have acquired the ability to make copies of themselves, or of each other (Figure 2).

One promising candidate for the role of the first biological macromolecule is today's RNA, because it can be both an information and an action molecule, as I will discuss in the section on ribozymes later. But some researchers believe that people searching for a universal genius, a single macromolecule that can be DNA, RNA, and proteins in one and self-replicate as well, are barking up the wrong tree. Stuart Kauffman of the Santa Fe Institute in New Mexico, for instance, likes to point out that not a single molecule in today's cells catalyzes its own replication. What really happens in a cell is that all molecular species are linked by a vast network of interactions including metabolism, synthesis of macromolecules, and eventually proliferation of cells. (If you walk around any biochemistry department in the world, you're likely to find a simplified version of this network on a huge wall chart produced by the biochemicals company Boehringer Mannheim.) Using computational analysis of theoretical model networks, Kauffman demonstrated that above a certain threshold of complexity and interactivity, such networks spontaneously evolve order from chaos. According to his views, one should not hunt for the ancestral molecule, but rather for the primeval metabolism.

From Building Blocks to Chain Molecules

"Odd, Watson—very odd!" said Sherlock Holmes. It certainly is odd when life arises from nothing on a previously lifeless planet, and the solution to this case that has dragged on for millions of years requires a detective's flair. Small wonder, then, that the Scottish chemist Alexander

 profile _____

Stanley Miller and the Primordial Soup

Working as a graduate student with Harold C. Urey (1893–1981) at the University of Chicago, Stanley Miller laid the foundation for the whole research field concerned with the origin of life. It was the year 1953, and Urey, who had received the 1934 Nobel prize in chemistry for the discovery of deuterium, the heavier isotope of hydrogen, was at that time mainly concerned with the chemistry of our atmosphere.

Miller constructed a simple laboratory apparatus designed to mimic the oceans, the atmosphere, and the thunderstorms of the primeval Earth. He kept the ocean boiling for a few days, while sending artificial lightning through the small-scale atmosphere containing methane, ammonia, hydrogen, and water vapor. Every once in a while he took a sample from the ocean to monitor its composition. After a few days, a reddish slime started to form, in which Miller found more than a dozen amino acids, including 6 of the 20 building blocks of today's proteins. He was also able to work out the reaction paths by which these organic compounds arise from inorganic gases, discovering intermediates like formaldehyde and hydrocyanic acid.

This surprising finding, made in the same year as the discovery of the double-helix structure of DNA, may have created the expectation that the missing steps between the primordial soup and the first cell (Figure 2) would soon be filled in. But today, some 40-odd years after his classic discovery, Miller is still puzzling over the same problem, and it doesn't look as if the solution is near. As yet, no one has been able to get the Urey–Miller experiment to produce molecules that might conceivably carry biological information or enzymatic activity. Therefore, most researchers have turned their back on simple models. More-evolved concepts often include solid substrates that may have served as scaffolds or catalysts for the first biological reactions. Even black smokers are being considered as possible sites for the original breeding place of life.

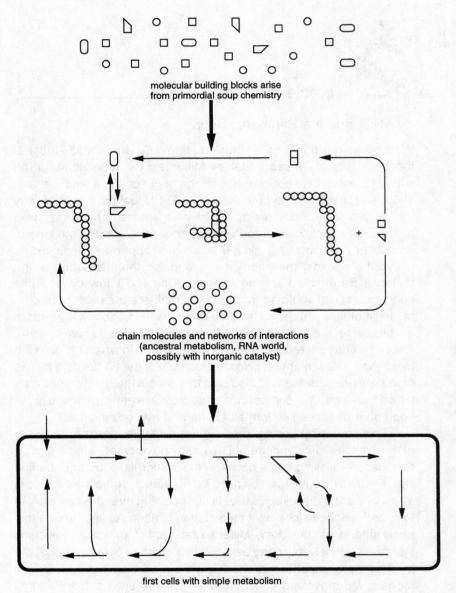

molecular building blocks arise
from primordial soup chemistry

chain molecules and networks of interactions
(ancestral metabolism, RNA world,
possibly with inorganic catalyst)

first cells with simple metabolism

Figure 2. From the primeval soup to the first cell. Schematic explanation of the fundamental steps of self-organization that were required for cellular life to come into existence.

G. Cairns-Smith asked the famous detective for help when he tried to solve this riddle.

This quote, from "The Naval Treaty," opens Cairns-Smith's book, *Seven Clues to the Origin of Life—A Scientific Detective Story*, in which he uses his detective qualities and a little help from Holmes to develop his hypothesis that life originated in the inorganic world. In his view, small irregularities in the crystal lattice of clay minerals were the first pieces of genetic information that proliferated when new layers crystallized with the same irregularities, including the possibility of minor mutations. After some time of inorganic evolution, a "genetic takeover" took place when the information-carrying clay minerals served as scaffolds for the first biological macromolecules based on organic building blocks.

Inorganic materials as a scaffolding for the first biomolecules also play a role in the competing theories of other scientists. Although nobody except Cairns-Smith goes as far as calling clay minerals the first life forms, various research groups have found that these and similar minerals can serve as catalysts speeding up the crucial step from small molecular building blocks to biological macromolecules. Although the British physicist J. D. Bernal made this suggestion as early as 1951, experimental data assigning such a role to clay minerals such as montmorillonite only started to accumulate in the 1970s. In the first impressive demonstration of the biological potential of clay minerals, M. Paecht-Horowitz, J. Berger, and A. Katchalsky of the Weizmann Institute of Science used the clay mineral illite to link amino acids coupled with an adenyl residue (as found in the energy carrier ATP) to a polypeptide. Apart from a single adenyl residue remaining at the end of the chain, the final product had exactly the same chemical structure as a protein.

Similarly, researchers were also able to synthetize nucleic acids using inorganic catalysts. In 1992, James Ferris and Gözen Ertem, working at the Rensselaer Polytechnic Institute in Troy, New York, managed to adsorb a variant of the nucleotide adenosine monophosphate (AMP) to a montmorillonite surface and polymerize it to form nucleic acid chains of up to 10 building blocks.

Quite a different inorganic compound is the cornerstone of a much talked about theory developed by the German patent attorney Günter Wächtershäuser, who started research into the origin of life as a hobby.[1] It is pyrite, an iron–sulfur compound widely known as "fool's gold" because of its golden glitter. Like Kauffman, Wächtershäuser does not believe in ancestral molecules, but rather in networks of chemical reactions. In his view, the ancestral metabolism may have derived its energy from the reaction of pyrite formation, while the product formed an ideal matrix to which simple organic molecules can bind reversibly. Minute pyrite crystals encapsulated by organic material may have been the precursors of the first cells. Even today there are still bacteria containing small crystals of iron sulfides and oxides. Some of them use their internal magnetite crystals for orientation relative to the Earth's magnetic field—they are therefore known as magnetotactic bacteria.

Although Wächtershäuser's theories have been partially confirmed by experiments carried out in the laboratories of Karl Otto Stetter in Regensburg, Germany, much of them remains inspired speculation for the time being. Both researchers agree in thinking that the origin of life happened in a rather hot environment. As I will explain, certain features of the family tree of life support this suggestion.

Another ancestral metabolism revolving around a different kind of sulfur compound has been suggested by the Belgian biochemist Christian de Duve, who received the 1974 Nobel prize for physiology/medicine for the discovery and characterization of certain cell organelles, the liposomes and peroxisomes. His theory, which he has presented in a popular science book, *The Origin of Life*, assigns a key role to a group of rather smelly chemicals, the thioesters. In fact, the sensibility of scientists' noses may be one of the reasons why these compounds have not yet been thoroughly tested in experimental model studies.

Ribozymes—Relics of a Lost World?

We know quite a few things about the last common ancestor of all organisms living today, a microbe known as the progenote. Essentially all the features that today's higher organisms have in common with bacteria, that is, most of the contents of a (nonmedical) biochemistry textbook, have been inherited from that ancestral cell. Thus we know that the progenote stored its genetic information as DNA and that it possessed a couple of hundred proteins working as enzymes, receptors, or transporters. The connection between information and function, between genes and proteins was the realm of RNAs (messenger, transfer, and ribosomal) as it is today.

However, it is important to stress that this common ancestor was already a quite complex biological system, evolved a long way since the birth of the first cells. Life's history beyond the progenote, the path from the first cell to the most recent common ancestor, remains in the darkness, because investigation of today's organisms cannot tell us anything about that part of our history. We can only speculate about how the first cells came into being, and which of all the compounds and functions that we now consider essential for cellular life may have been absent in the prototypes of the cell. The main constraints guiding our thought are the conditions of the primordial soup as a starting point and reservoir of molecular building blocks, and the progenote as the final product. (Note that there may have been hundreds of other species contemporary to the progenote, whose descendants just didn't have the fitness to survive into our time.)

One of the most popular theories concerning the phase between primordial soup and ancestral cells is the RNA-world hypothesis, which received a boost in the 1980s when Sidney Altman and Thomas R. Cech discovered that RNA can possess catalytic activity. Before today's distribution of tasks between DNA, RNA, and proteins came into being, so the theory goes, RNA may have been the only biological macromolecule and may have carried out simpler versions of the tasks of all three kinds of macromolecules found in modern cells. The newly discovered RNA catalysts were called ribozymes, and their discoverers awarded the Nobel prize for chemistry in 1989.

The discovery of ribozymes as potential relics of the RNA-world was a major inspiration for RNA biochemists, who set out to find or make RNA molecules with the ability to carry out some of the functions considered essential for life (see Focus, "Ribozymes with New Activities and New Structures," for further details). However, if the goal was to reconstruct the RNA-world, success has been rather modest. Although almost everyone agrees that the ribosome, the key element in the translation of the nucleic acid code into proteins, most probably evolved from a catalytic RNA molecule, no one has been able to reconstruct a functional RNA-only ribosome. And the exact details of the ribosome's atomic structure have not been solved either, despite much effort.

Archaebacteria: A New, Very Old Domain of Life

The emergence of the first cells marked the beginning of the longest era in the history of life on Earth, the era of the unicellular organisms. Only

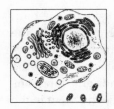 focus _____

Ribozymes with New Activities and New Structures

One reason why the RNA-world has remained a controversial hypothesis is that the catalytic functions observed by Cech and Altman were quite primitive. For a decade, the proven abilities of ribozymes were limited to self-splicing and similar RNA–RNA interactions. Only in the mid-1990s has the list of ribozyme-catalyzed reactions been extended substantially.

The first ribozyme-catalyzed reaction that does *not* occur on the phosphate group of the RNA backbone was observed by Cech's group in 1992 with a genetically modified *Tetrahymena* ribozyme. The altered ribozyme could cleave the ester bond between the amino acid methionine and the matching tRNA, so it did the reverse of what tRNA synthetases do in the cell, establishing a first connection between ribozymes and protein biosynthesis. Three years later, Michael Yarus, a colleague of Cech at the University of Colorado, and his group screened a random mixture of 10^{14} different RNA sequences and managed to select a sequence that catalyzes the attachment of an activated amino acid to its tail. The reaction was 10,000 times faster than with a noncatalytic RNA. Similarly, Peter Lohse and Jack Szostak at Harvard selected an RNA species able to transfer aminoacyl residues as the peptidyl transferase center of the ribosome does when it elongates the polypeptide chain by one unit. Together with the finding of Harry Noller's group at the University of California, Santa Cruz, that none of the ribosomal proteins is essential for peptidyl transferase activity, this work boosted the view of the ribosome as a catalytic RNA stabilized and supported by proteins.

Until very recently, the bottleneck on the road to a better understanding of ribozyme function was the lack of structural data. Although the base pairing can be predicted from the sequence complementarity with relative ease, the treelike secondary structure diagrams obtained from this analysis cannot tell us anything about the three-dimensional structure. For instance, the simple cloverleaf diagram of transfer RNAs does not contain the information that the actual molecule has the shape of an L.

The first catalytic RNA to have its structure solved was the hammerhead ribozyme, which cuts into the proper pieces the multicopy string resulting from the replication of the genome of the avocado-sunblotch virus. Using different approaches to avoid self-cleavage during the structure determination, the groups of David McKay at Stanford and Aaron Klug at the University of Cambridge arrived at essentially identical structures, revealing a Y shape with a binding site for magnesium ions close to the cleavage site. From the structural data William Scott and Aaron Klug derived a hypothetical description of the catalytic mechanism of this ribozyme.

The ribozyme–inhibitor complexes used in the hammerhead studies only contained 34 nucleotides (half a tRNA). However, many catalytically active RNA introns are many times larger than tRNA, to say nothing about the ribosomal RNAs. In 1996, the groups of Thomas Cech and Jennifer Doudna solved the first structure of an RNA molecule substantially bigger than tRNA, namely the 160-nucleotide-long P4–P6 domain of the *Tetrahymena* ribozyme. The domain, which contains half of the ribozyme's active center, turned out to be surprisingly compact in its structure. This is made possible by the presence of numerous magnesium ions that balance the negative charges of the nucleic acid and thus suppress repulsive forces. A number of structural features that may be common in big RNAs were observed for the first time in this study, including the parallel packing of RNA helices.

The demonstration that structures of large RNAs can be solved sparked discussions of the folding properties of RNA in analogy to protein folding, including the possibility of RNA chaperones. The (RNA-)world is looking forward to further big structures, like the complete *Tetrahymena* ribozyme, and—why not?—the ribosome. A deeper understanding of today's catalytically active RNAs might be the key to the history of life before the progenote.

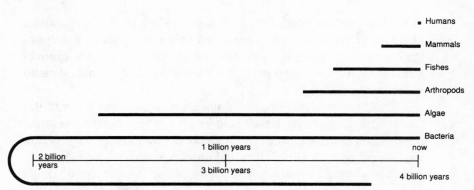

Figure 3. The clock of the development of life on Earth. During five-sixths of the time that has passed since the origin of life, our planet was populated exclusively by single-cell microbes.

600 million years before our time, multicellular organisms came into being. Thus, for a span of three billion years, or more than two-thirds of its history, our planet was populated exclusively by microbes (Figure 3). In this perspective, the whole evolution of animals and plants appears like an afterthought, the history of humankind a mere moment. What happened during these three billion years is not recorded in fossils like the later evolution of higher organisms. Although there are microfossils from this time, the information they provide—mainly about the outer shape and size of the microbes concerned—does not help to sort out the kinship between microbial species and families. Whereas fossils of higher organisms reveal complex properties such as skeletal architecture, which are controlled by many genes and thus are quite reliable indicators of heredity and kinship, the shapes of bacteria only depend on very few genes and can vary quite easily among a limited set of possibilities. Thus, one particular shape of a microbial cell may have been "reinvented" by evolution several times in completely unrelated organisms.

Despite this handicap, microbiologists now have a realistic chance of obtaining as much information on the time of bacterial dominance as we have on the (relatively recent) time of the dinosaurs. All creatures living today are related to each other; they are all descendants of one common ancestor (the progenote), whose genes have changed in many different ways in the different lineages of the family tree. In many essential areas, however, they have remained sufficiently similar to allow scientists to make comparisons and compute degrees of kinship and the time elapsed since the separation of lineages. From its painful beginnings in the early 1970s to this day, this kind of molecular paleontology has come up with many surprises. All too often, biologists classifying living things by their outer appearance had been misled. The most spectacular rearrangement was the introduction of a new, third domain of life, which, in a sense, is also a very very old domain of life. Its members are known as archaebacteria, or archaea, for short.

With the help of a microscope, one can divide living cells into two kinds. There are big ones (typically 10 micrometers in diameter), which have various compartments, including a cell nucleus surrounded by a membrane. And there are smaller ones (typically 1 micrometer in diameter), which are more simply structured, contain less DNA, and do not contain any subcompartments (Figure 4). The first type, the eukaryotic cell, is the building block of all multicellular organisms including ourselves, but it also occurs as a free-living microbe, with the baker's yeast *Saccharomyces cerevisiae* as the classic example. The small and simple type, known as the prokaryotic cell, occurs only in unicellular organisms.

As was just explained, the family tree of prokaryotes cannot be derived from the fossil record, but has to be looked for in the genetic inheritance of today's bacteria. Proteins were the first kind of biological macromolecule whose information content could be read, following the sequencing of insulin by Frederick Sanger.[2] Protein sequencing has indeed been used for evolutionary analysis, but it has several shortcomings. For instance, not every protein is present in every organism. Furthermore, similarity of proteins in diverging lineages may increase instead of decreasing as expected. If, for example, two unrelated proteins carry out similar functions in different organisms, they may develop similarities dictated by their function rather than by kinship. This effect, known as convergent evolution, can also be observed on the anatomical level in many examples from higher organisms, such as the finlike appearance of the penguin's wings.

Therefore, when sequencing short fragments of DNA and RNA became feasible at the end of the 1960s, some researchers preferred to study evolutionary kinship on the gene level. As there are four DNA bases grouped in "words" (codons) of three letters, allowing $4^3 = 64$ genetic

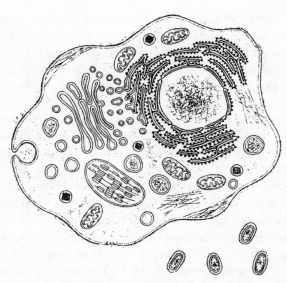

Figure 4. Comparison of eukaryotic and prokaryotic (bacterial) cells. Typical eukaryotic cells are tenfold greater in each dimension than bacteria (so they contain
1,000 times the volume!). They have various subcellular compartments with specialized functions. Some of these organelles are thought to be the descendants of
bacteria that lived in a symbiotic relationship with the primeval eukaryotic species.

words to specify just 20 amino acids, there is a certain amount of random
mutation possible on the gene level that cannot be affected by convergent
evolution.

However, when Carl Woese started such investigations at the University of Illinois in 1969, he did not even use protein genes, but the sequence
of one molecule that he knew every cell possesses, namely the RNA of the
small ribosomal subunit, called 16S RNA in bacteria (S for Svedberg units,
in terms of which the speed of sedimentation of a particle is measured).
With some 1,500 nucleotide "letters," this molecule is quite big enough to
allow reliable statistical analysis and to avoid chance similarities. However, in the early 1970s, sequencing a gene or an RNA molecule of this size
in one go was unthinkable for one organism, let alone for scores of them
for evolutionary studies. Therefore, Woese cut down the molecule into
conveniently sized fragments by digesting it with an enzyme that cleaves
the strand after each letter G. Thus he ended up with a lot of words of

different length, but all ending with a G, such as AUG, CACUAAUCAG, CG, UUUAAG, and so on.

He found that the set of six-letter words provided the ideal material for statistical analysis of the evolutionary relationships among bacteria, so he set out to record RNA fingerprints of as many bacterial species as possible. By 1981 (that is, before the sequencing of nucleic acids became much easier by progress in molecular biology methods including the polymerase chain reaction; see Chapter 4), he and his co-workers had characterized more than 200 species. When the researchers began drawing family trees, they had a big surprise. The group of bacteria producing methane (CH_4) out of carbon dioxide (CO_2) did not fit into the family tree of the other bacteria. Before the bacteria split into the lineages leading to today's families, genera, and species, the methanogens had gone their separate way. Within their group, the split between species is as deep as within the normal bacteria. In some respects, the methanogens are more ancient than other bacteria. For instance, they can only live anaerobically, that is, in the strict absence of oxygen, implying that they failed to change their habits when the Earth's atmosphere changed from a reducing to an oxygen-rich, oxidizing one.

During further studies, two other large groups were found to join the methanogens, namely the thermoacidophilic and the halophilic bacteria. On the basis of his RNA analysis, Woese drafted a new family tree of life, which instead of the classic two branches (eukaryotes and prokaryotes), had three (Figure 5). The new domain, which he called archaebacteria (and renamed archaea later), was to stand side by side with the "true" bacteria (eubacteria or just bacteria, now defined as all prokaryotes that are not archaebacteria) and the eukaryotes, which still stand apart for having a nucleus and various other organelles.

Although Woese's hypothesis met with a lot of disbelief among microbiologists, everything that reseachers found out about the archaebacteria—most of which are extremophilic and "eccentric" in other respects as well—confirmed the notion that they are a world apart. For instance, the analysis of their cell membranes revealed molecular building blocks that no one had dreamed of before. This suggests that the split between archaebacteria and all other living things even occurred at a time when the membrane of the ancestral cells was not yet fully developed. Still, some scientists remained skeptical, until in 1996 the complete genome sequence of an archaebacterium was published.

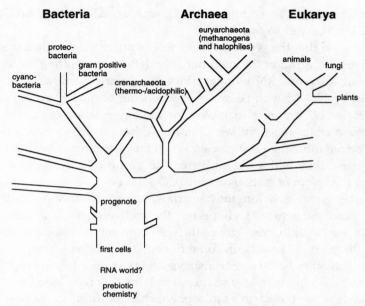

Figure 5. The tripartite family tree of life according to Woese. There is still some debate as to how the three main branches are to be positioned with respect to each other and to the root of the tree.

Methanococcus jannaschii: Decoding an Archaebacterium

Based only on the analysis of short fragments of ribosomal RNA, Carl Woese put forward the revolutionary hypothesis of archaebacteria as a third domain of life. Some 20 years after this painstaking effort to reveal the family tree of life, the sequencing of genes has become a routine work only limited by the funds available for it; by now the sequences of entire genomes of microbial organisms are being decoded more rapidly than anticipated. Following two bacteria (*Haemophilus influenzae* and *Myco-plasma genitalium*) and the (eukaryotic) baker's yeast *Saccharomyces cere-visiae*, the archaebacterium *Methanococcus jannaschii* became in 1996 the fourth organism to be sequenced in its entirety. (Although the *Escherichia coli* sequencing project was started first and this bacterium was initially the best characterized, it was only fifth in having its sequence published; see

Table 1. *Helicobacter pylori*, the ulcer-causing bacterium mentioned in Chapter 4, has been sequenced by industry researchers, who did not publish the data, and then resequenced and published in 1997.) Thus, with at least one representative of each of the three domains sequenced and dozens of sequences expected to come out soon, Woese's classification can now be tested by the comparison of whole genomes.

The sequencing of the methanogenic archaebacterium, which was discovered at a black smoker location in the Pacific in 1982 and named after the Woods Hole microbiologist Holger Jannasch, completed a spectacular hat-trick for a scientist who had started as an outsider in the genome race. The laboratory of Craig Venter at the Institute for Genomic Research in Rockville, Maryland, had pioneered the method called shotgun sequencing, in which huge amounts of DNA, such as the entire genome of *Haemophilus influenzae* with all its 1.8 million nucleotides, are cut down into thousands of random fragments. The researchers only sequence the ends of these pieces and then use massive computing power to try to piece together large bits of the genome from overlapping fragments (with a miniumum of 50 identical nucleotides), and thus eventually reconstruct the whole genome (Figure 6). Once they have identified the fragments needed for this reconstruction, they sequence the remaining parts of these fragments only. In contrast, conventional sequencing starts with identifying genes, drawing a gene map, followed by selecting and sequencing the genes one by one. Venter's alternative approach was not taken seriously in the beginning. For instance, the National Institutes of Health refused to fund his project to sequence the genome of *Haemophilus influenzae*, which he nevertheless completed in just 13 months. This was the first genome sequence ever, and just four months afterward, Venter's institute also made second place with *Mycoplasma genitalium*.

Forty researchers working at Venter's institute and at five other laboratories are listed as authors of the paper reporting the sequencing of the 1.7 million-nucleotide genome of *Methanococcus jannaschii*. A total of 36,718 fragments were sequenced and pieced together to represent the entire ring-shaped chromosome of the archaeum plus two small extrachromosomal elements (ECEs). Judging by the occurrence of the typical start and stop sequences, the chromosome contains 1,682 genes coding for proteins, and the ECEs a further 44 and 12, respectively.

The preliminary comparisons of the 1,738 protein genes of *Methanococcus* with the known sequences of other organisms yielded a triumphant

Table 1. The First Full Genome Sequences Published

Organism (domain)[a]	Genome size (million basepairs)	Institution[b]	Publication year
As of 31 July 1997			
Haemophilus influenzae (b)	1.83	TIGR	1995
Mycoplasma genitalium (b)	0.58	TIGR	1995
Methanococcus jannaschii (a)	1.66	TIGR	1996
Synechocystis (b)	3.57	KDRI	1996
Mycoplasma pneumoniae (b)	0.81	University of Heidelberg	1996
<u>*Saccharomyces cerevisiae*</u> (e)	13	International consortium	1997
<u>*Escherichia coli*</u> (b)	4.60	University of Wisconsin	1997
<u>*Helicobacter pylori*</u> (b)	1.66	TIGR	1997

. . . and the queue[c]

Bacteria
 Borrelia burgdorferi (1997)
 <u>*Deinococcus radiodurans*</u> (1997)
 Streptococcus pneumoniae (1997)
 Treponema pallidum (1997)
 Ureaplasma urealyticum (1997)
 <u>*Vibrio cholerae*</u> (1998)
 <u>*Thermotoga maritima*</u> (1998)
 Actinobacillus actino- mycetemcomitans
 <u>*Aquifex aeolicus*</u>
 <u>*Bacillus subtilis*</u>

 Caulobacter crescentus
 Chlamydia trachomatis
 Enterococcus faecalis
 Legionella pneumophila
 Mycobacterium avium
 Mycobacterium tuberculosis
 Neisseria gonnorrhoea
 Neisseria meningitidis
 Porphryomonas gingivalis
 Pseudomonas aeruginosa
 Rickettsia prowazekii
 Salmonella typhimurium
 Shewanella putrefaciens
 Streptococcus pyogenes
 Treponema denticola

Archaea
 Archaeoglobus fulgidus (1997)
 Halobacterium salinarium
 Methanobacterium thermoautotrophicum
 Pyrobaculum aerophilum
 <u>*Pyrococcus furiosus*</u>
 Pyrococcus shinkaj
 <u>*Sulfolobus solfataricus*</u>
 <u>*Thermoplasma acidophilum*</u>

Eukarya
 Plasmodium falciparum (1997)

[a]The names of organisms mentioned elsewhere in this book are underlined.
[b]TIGR, The Institute for Genomic Research, Rockville, Maryland; KDRI, Kazusa DNA Research Institute, Chiba, Japan.
[c]Source: TIGR website (http://www.tigr.org./). Genomes expected to be published by the end of 1998 are listed first, followed by an alphabetical list of the microbial genomes currently in progress.

confirmation of Woese's family tree with three domains. Surprisingly, 56 percent of the genes could not be assigned to any related genes in bacteria or eukaryotes. In contrast, when *Haemophilus influenzae* was sequenced, 80 percent of the genes had counterparts in existing databases.

Obviously, the high number of novel genes in *Methanococcus* is a potential bonanza for microbial geneticists. Knowing the sequences, they can knock out the genes individually and try to work out their function from the consequences of their deletion. It is a safe bet that they will discover novel proteins, including some with potential applications in biotechnology or medicine. One hitherto small research field that has received a major boost from the *Methanococcus* sequence is the study of self-splicing proteins (see Focus, "Inteins Everywhere").

While the low percentage of homology with sequence data in databases (mainly populated by bacteria and eukaryotes) already confirmed the existence of the third domain, comparisons of the genes with known function proved the point even more clearly. Dividing the genes into groups corresponding to certain tasks in the cell, it is found that some groups of archaebacterial genes are amazingly similar to their eukaryotic counterparts, while others are more like the corresponding eubacterial genes. For instance, archaea resemble higher organisms in the genetic setup of their information technology. The molecular machinery essential for copying and translating DNA and RNA has pretty much the same cogs and wheels in methanogens as in yeast, mice, and men, but differs markedly from that of bacteria. In contrast, most of the genes having to do with basic metabolic reactions reveal that archaebacteria have kept the bacterial lifestyle, and have little in common with eukaryotic metabolism.

The group of genes responsible for the methane production, of course, resembles nothing but other methanogens, and these are all members of the third domain. Geneticists now believe that they have identified all the genes involved in this very special branch of metabolism. In addition, *M. jannaschii* is able to fix nitrogen (i.e., convert the inert atmospheric gas into something biochemically useful). The genes required for this task have also been identified in the genome.

Sometimes the absence of a gene can be an interesting finding as well. With the availability of complete genomes, it is now possible for the first time to state with confidence that a certain gene is absent in a given organism. And in this respect, too, the methanogen has some remarkable properties. For instance, it does not possess a charging enzyme (tRNA

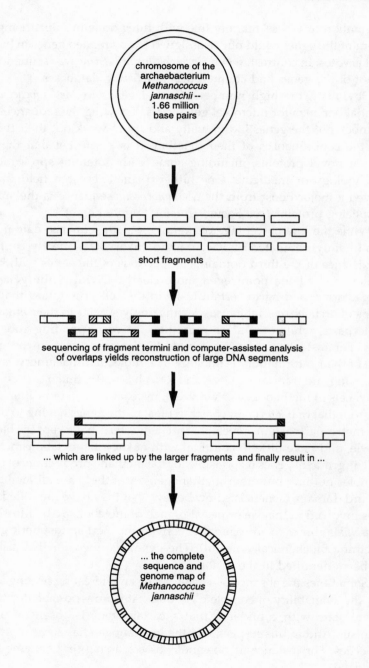

chromosome of the archaebacterium *Methanococcus jannaschii* -- 1.66 million base pairs

short fragments

sequencing of fragment termini and computer-assisted analysis of overlaps yields reconstruction of large DNA segments

... which are linked up by the larger fragments and finally result in ...

... the complete sequence and genome map of *Methanococcus jannaschii* synthetase) for each of the 20 amino acids to be bound specifically to the correct tRNA. Scientists suspect that the amino acids lacking such an enzyme are only generated from related amino acids (e.g., glutamine from glutamic acids) after binding to the tRNA.

Of course, there is a limit to the conclusions that one can draw on the basis of one or two representatives of each domain. It is certain, however, that genome sequencing has removed the last traces of reasonable doubt concerning the tripartite family tree of life. Forthcoming genome sequences of other organisms will improve the datasets for evolutionary studies. Eventually, genome sequencing might help to answer one of the most intractable mysteries of molecular biology: the origin of the eukaryotic cell.

Do We All Come out of the Heat?

Many scientists consider archaebacteria as a window into the history of life on our planet. Their peculiar preferences for extreme conditions may be relics of the adaptation to conditions that were prevalent some three billion years before our time. The observation that hyperthermophilic adaptation is particularly widespread among archaebacteria led some researchers to believe that adaptation to extreme conditions is not a feature

←——————————————————————————————————————

Figure 6. Schematic explanation of the "gunshot" strategy employed by Craig Venter's institute for the sequencing of complete prokaryotic genomes, including the first archaebacterial one, *Methanococcus jannaschii*. Random fragments of the genome are organized into two libraries: one with many small fragments that can be cloned in plasmids; the other with fewer, longer DNA strands, which are cloned in the bacteriophage lambda. In the case of *M. jannaschii* the first library contained 36,000 fragments that had 2,500 base pairs or less. These random fragments cover each base of the genome 50 times on average, so one does not have to sequence them all. Only a part of the fragments is sequenced from both ends until there is enough overlap to allow the reconstruction of major parts of the genome, the so-called contigs. For *M. jannaschii*, 17,000 fragments were partially sequenced, which allowed a computer-assisted assembly of 14 contigs. Researchers then used the second library to link these contigs and close any gaps that may have been between them. Partial sequencing of 337 larger fragments finally completed the reconstruction of the complete ring-shaped chromosome of the archaebacterium. In the end, the researchers only had to complete the sequencing of those fragments that were part of the chain of overlapping fragments representing the genome.

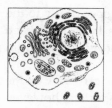 focus _____

Inteins Everywhere—a Surprising By-product of the *Methanococcus* Sequencing

Inteins—self-splicing proteins—used to be considered as rather exotic, until 1996, that is. Then, in the sequence of the *Methanococcus* genome, 18 such sequences were discovered in 14 different genes, which more than doubled the number of known examples. In analogy to RNA introns, inteins can cut themselves free from a longer polypeptide chain and link the remaining bits (called exteins in analogy with RNA exons). This process, however, was only discovered in 1990 and has not received as much attention as RNA splicing. Although most of the known examples of inteins come from the domain of the notoriously idiosyncratic archaebacteria, the prototype was discovered in a well-studied organism that has been of service to humankind for millennia: *Saccharomyces cerevisiae* (baker's yeast). Tom H. Stevens and his coworkers at the University of Oregon at Eugene observed that the *TFP1* gene of yeast codes for two protein products. The smaller one is coded in the middle region of the gene and flanked by the separate halves of the bigger one. This finding on its own would not have been very remarkable. There are many examples of overlapping or nested genes. It was unusual, however, that instead of the expected two messenger RNAs (one for each protein product) only one was found, and its length corresponded to the sum of the lengths expected. Stevens' group therefore suspected that the genetic information is not, as in many other cases, edited on the mRNA level. Rather, the single mRNA seemed to get translated into a single fusion protein, which splits into the final two components after translation.

To test this hypothesis, the researchers generated mutations in the middle part of the mRNA, leading to a shift of the reading frame, i.e., to a wrong segmentation of the string of nucleotides ("letters") into three-letter words specifying the amino acids to be incorporated into the protein. This kind of mutation not only affects the word where it occurs—the whole text behind it will be distorted as well. If there had been a splicing on the mRNA level, the frameshift would have only affected the intron cut out of the mRNA, as the spliced exons should still have had the correct frame. Therefore, only the protein coded by the middle segment should be mutated, not the one coded by the outer parts. It was found, however, that both proteins were affected by the frameshift. (Of course, one has to take care that the mutations of the middle part do not affect the splicing reaction, as can be confirmed by the molecular weights of the two products.) However, the researchers did not succeed in isolating the uncleaved precursor protein. This led them to suspect that the splicing might be an autocatalytic process. In this case, the protein itself would make the splicing reaction occur so rapidly that it would be impossible to get hold of the original translation product.

This difficulty was only overcome when inteins were also discovered in several hyperthermophilic archaebacteria, such as *Thermococcus litoralis* and various species of *Pyrococcus*. Francine B. Perler and her co-workers at New England Biolabs constructed an artificial self-splicing system around the intein (I) of *Pyrococcus* DNA polymerase by putting the gene for a maltose binding protein (M) in front of it (a so-called N-extein, since it is at the amino-, or N-terminal end of the sequence) and a paramyosin gene (P) behind it (a C-extein, for carboxy-terminal). They introduced this fusion gene into the intestinal bacterium *Escherichia coli*, whose protein synthesis apparatus duly made the fused polypeptide (MIP) at temperatures between 12 and 32°C. At these low temperatures, the self-splicing reaction occurred only very slowly, as the intein involved came from an organism adapted to life near the boiling point of water. In fact, the whole process was slowed down to such an extent that the researchers were able to purify the unprocessed precursor protein. Incubating this polypeptide in aqueous solutions containing only small amounts of sodium chloride and phosphate buffer and then warming it slowly, they could observe the onset of the self-splicing at higher temperatures. This way, they could also isolate an intermediate (MIP*) that behaved rather paradoxically. It appeared to have a higher molecular weight than MIP, as it moved more slowly through electrophoretic gels, and it also seemed to possess two different amino termini. The riddle was solved by the finding that the intermediate obviously has a branched structure, whose bulkiness decreased the mobility in gels. What must have happened is that the exteins M and P already formed a link while I was still attached to P.

In addition to their self-splicing abilities, inteins share a further characteristic with introns—or, more specifically, with the proteins derived from intron translation. Both act as endonucleases, which means that they can recognize certain DNA sequences and cut the DNA at a well-defined position within or near these sequences. They are specialized on DNA segments characteristic of their "home" gene, but lacking the intein or intron sequence. Thus, they cut the gene at the corresponding position and thereby trigger a "repair" mechanism that may use an intein/intron-containing copy of the gene as a template and thus produces the intervening sequence and inserts it into the gene. This process—traditionally known as "intron homing"—has thus far only been demonstrated for four of the known inteins. Sequence comparisons, however, allow the conclusion that all known inteins are at least related to endonucleases, even if some of them may have lost the homing function in their recent evolution. The sporadic distribution of intein sequences among species of all three domains of life also suggests that this endonuclease activity has facilitated the spreading of inteins through horizontal gene transfer (i.e., transfer between contemporary species).

Why have self-splicing inteins and introns developed the function of homing endonucleases? Well, this question is put the wrong way around. If you turn it around, the answer is almost self-evident. Why are homing endonucleases self-splicing, be it on an RNA or on a protein level? An enzyme that is able to cut a gene and put its own genetic material right in the middle of it is potentially lethal for any cell, as it will at some time destroy an essential gene. If, however, the damage can be repaired on the RNA or on the protein level by means of the inserted sequence catalyzing a splicing reaction reestablishing the original product, all is well.

Thus, the intein or intron activity is the condition that a certain kind of mobile genetic element has to fulfill in order to be tolerated by the cell. If they don't fulfill it, they are damaging their own host cell and have only one chance of surviving—to acquire the ability to infect other cells and thus become viruses. Thus it is most probably no coincidence that self-editing on the protein level is most commonly observed in viruses. The genetic material of the AIDS-causing virus, HIV, for instance, codes for a long polypeptide chain containing all the viral proteins linked together. The HIV protease activity contained in this polyprotein cleaves the molecule into the desired proteins.

Perhaps the biological "purpose" of the inteins (which are obviously useless for the cell and only survive because they can help the spreading of their gene without hurting the cell too much) is to be understood as an evolutionary precursor or a peaceful alternative to viruses. One should definitely keep an eye on them.

---

acquired during the course of evolution, but one that most of today's organisms have lost in the course of an evolution away from hyperthermophilic ancestors.

Evidence for this hypothesis comes not only from the archaebacterial domain. If one screens the entire tree of life for thermophiles, one tends to find them in the deepest branches, i.e., in those groups that separated relatively early from the mainstream. The order Thermotogales (including *Thermotoga maritima* mentioned earlier), for instance, is the only hyperthermophilic group in the domain of eubacteria, and at the same time, it is the deepest branch of this domain (Figure 7).

While these findings point to hyperthermophilic ancestors, one has to be careful to keep in mind that they are not evidence for a "hot" origin of life, because conclusions drawn from the family tree of today's species can only reach back to the progenote (common ancestor), as I explained earlier. Even if one accepts the conclusion that the progenote was thermophilic, this does not imply that the very first cells lived in the heat as well. There may have been nonthermophilic species contemporary with the progenote that have died out. To illustrate this point, let us imagine a realistic scenario assuming that at the time of the progenote, there was a reasonable degree of biodiversity in the oceans, with hundreds of microbial species each adapted to different environmental conditions (including temperatures) and lifestyles. At the time, the last major meteorite impact may have brought much of the ocean water to a boil, with the effect that only one thermophilic species survived the heat shock to become the ancestor of all

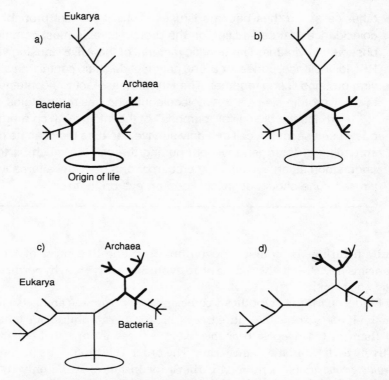

Figure 7. Different hypotheses concerning the placement of the hyperthermo-philes in the family tree of life. (a) "Hot origins" of life. (b) Development of hyperthermophilic properties before the split into the three domains. (c) Development of hyperthermophiles after the split between eukaryotes and prokaryotes. (d) Independent developments leading to hyperthermophiles in various branches of eubacteria and archaebacteria.

present-day life on Earth. When climatic conditions returned to normal, some of its descendants would have lost the heat resistance as they had no use for it in their everyday lives, while others retreated to extreme environments where the thermophilic adaptation provided them with exclusive access to an ecological niche. Thus, to be brutally honest, we still do not know whether the very first cells lived in lukewarm brine like *Escherichia coli* or in boiling water like *Pyrococcus furiosus*. And perhaps we will never find out.

Searching for Gaia: Life on Earth as a Hyperorganism

I mentioned that the climate of our planet has remained amazingly constant over the past three billion years, considering how easily the chaotic interactions between celestial bodies could have driven Earth into the abyss of sterility. It is not very difficult to make up scenarios of catastrophes that might have happened. To quote just two grossly over-simplified examples showing that life could have died by overheating as well as by freezing:

- Chemical reactions could have bound a greenhouse gas from the atmosphere in minerals (carbon dioxide to carbonate, for instance). This could have led to a drop in the average temperature by just a few degrees, which could have increased the size of the polar ice caps and thus the fraction of sunlight reflected into space unused (termed the albedo, i.e., "whiteness" of a planet). Positive feedback among a reduced energy intake, further drop in temperatures, and further growth of the ice caps could have led to a planet in the deep freeze, like our neighbor Mars.
- Conversely, we could have ended up in a planetary oven, considering that the luminosity of the Sun has increased by some 40 percent over the past three billion years. This would have been more than enough to send us on an overheating feedback loop, ending in a state comparable to that of our other neighbor, Venus.

We owe our sheer existence to the astonishing fact that our planet remained within the temperature range in which water is liquid on the major part of its surface for three billion years, despite quite important changes in the energy transferred from the sun and in the composition of the atmosphere. This observation led the British chemist James Lovelock (see his Profile) to suggest that a well-tuned mechanism of self-regulation must be at work. He postulated that life on Earth actively kept the conditions roughly constant by the use of negative feedback mechanisms, similar to the way in which we and some other animals keep our body temperature constant even if the environmental temperature changes drastically.

This hypothesis was already quite heretical in its own right, and Lovelock further provoked the anger of skeptical colleagues by naming the cybernetic system of geosphere and biosphere after the Greek Earth

profile _____

James Lovelock—a Heretic?

The contempt of the science establishment was unanimous: The ideas in question were bare nonsense, even "a danger for science." Only the reviewer writing for the top class science journal *Nature*, normally the guardian of science against all kinds of dangers, found words of understanding for the allegedly dangerous biologist Rupert Sheldrake, whose ideas about "morphic fields"—claimed to enable telepathic learning—aroused much public interest, but were shunned by the scientific community. The reviewer writing for *Nature*, however, argued that the only danger for science came from those colleagues who invested much fundamentalist zeal into censoring everything not confirming the one and only doctrine, and did not acknowledge the possibility of fruitful errors. These were not the words of somebody who oscillated in morphic resonance with Sheldrake himself, but of a fellow sufferer, of someone whose theories had also been called "a danger for science" at one point. The judicious reviewer was none other than the British chemist and inventor James Ephraim Lovelock, FRS.

In the 1960s, Lovelock had invented new methods for the detection of extremely small concentrations of certain gases in the atmosphere. He could for the first time measure the accumulation and spread of the halogenated hydrocarbons and thus laid the foundations for the work of Mario Molina, F. Sherwood Rowland, and Paul Crutzen on the threat of these substances to the ozone layer, which was rewarded with the Nobel prize in chemistry in 1995. This invention, along with his research in cryobiology, transfer of infection, gas chromatography, and atmospheric chemistry, earned Lovelock a high profile, and he became a Fellow of the Royal Society in 1974.

However, once he had reached the inner circles of the science establishment, Lovelock, who calls himself a nonconformist, must have had an upsurge of revolutionary spirit. He moved the focus of his research away from analytical atmospheric chemistry, where he had been an expert respected worldwide, to the verification of his pet theory, which initially brought him nothing but trouble. The notion of the Earth as a hyperorganism was called unscientific, new-age mysticism, and worse. Lovelock retired from institutionalized science and carried on stating his arguments from his home in rural Cornwall.

Gaia, however, withstood her attackers and left the battlefield unbeaten. Some of the global self-regulating mechanisms proposed by Lovelock hypothetically in the 1970s were confirmed by data measured in the 1980s. The Gaia hypothesis evolved into the Gaia theory.

Nowadays, Lovelock's Gaia theory is almost generally accepted— at least as a reasonable hypothesis. In spring 1994, for instance, a major international conference about Gaia was held at Oxford, where prominent supporters (like the biologist Lynne Margulis) and skeptics (like the evolutionary theoretician John Maynard Smith) exchanged their arguments. Nobody calls for censorship anymore when Gaia is mentioned, and Lovelock has received major awards, including the Volvo environment award in 1996 and the Blue Planet Prize of the Asahi Glass Foundation in 1997.

---

goddess Gaia, and speaking of Gaia as "she," as of a living female creature. The fact that he presented his ideas in a popular science book rather than in a scientific journal did not make him more popular with the science establishment, either. At one point, he was quite close to suffering the modern multimedia equivalent of being burned at the stake.

Luckily for Lovelock and his Gaia theory, some of the global feedback mechanisms that he had postulated out of the blue could later be demonstrated with solid data. Even if the name Gaia and the female personal pronoun are still shocking to some scientists, "geophysiology" has become a respectable field of research. What happens on a global scale if the ozone hole increases or if the flowering of overfed algae becomes a natural catastrophe are the sorts of questions that Lovelock has taught us to answer.

But what role do extremophiles play in the metabolism of the Earth mother Gaia? Not a very important one, says Lovelock in his book *Gaia—*

A New Look at Life on Earth. He compares the extremophilic microbes with eccentrics, who can only be tolerated by an affluent society. I disagree with this statement in both parts—neither extremophiles nor human eccentrics are luxury items. Both our society and the biosphere need a wide range of diversity. In the scenario concerning the heat adaptation of the progenote I demonstrated that the existence of eccentric heat-resistant microbes may at some point have saved the existence of life on Earth.

Apart from the new insights that the Gaia theory gave us about the history of Earth and its biosphere, the global viewpoint of the planet as a cybernetic system pioneered by Lovelock is also useful for the discussion of the possibility of life on other planets, which we will enter in the next chapter.

Endnotes

1. Although the science establishment tends to mistrust hobby scientists (and more generally anyone spending less than 80 hours per week in a laboratory), some have already noted that Wächtershäuser is not the first German scientist dealing with patents in his day job while making up theories in his spare time. That's exactly what this other guy did, the "third class technical expert" of the Bern patent office, what was his name again?—oh, yes, Albert Einstein . . .
2. Frederick Sanger, born 1918, British biochemist, received the 1956 Nobel prize for chemistry for the first sequencing of a protein. In 1980, he was awarded the prize again, together with W. Gilbert and P. Berg, for contributions to the sequence analysis of nucleic acids.

Updates

p. 145   The complete genome sequence of *Thermotoga maritima* published in 1999 shows massive horizontal gene transfer even between the domains of eubacteria and archaea, explaining why family trees based on small numbers of genes are often contradictory. Analysis of the genome sequence of *Pyrococcus abyssi* has revealed that this archaeon replicates its DNA in a manner very similar to that of bacteria, even though the enzymes involved are more closely related to those in eukarya.

p. 147   More recently, the view that our climate was not quite so constant and that life on Earth may have been very close to the edge of that abyss has won more and more support. See "Escape from Frozen Hell" in the Afterword.

6

Life beyond Earth

We have now explored the limits and the extreme margins of life on our planet from freezing to boiling, into the deepest trenches of the oceans, and from the origin of life to this day. How about elsewhere? you may now ask. Are these the limits and extremes of life in general? Whether there is life "out there," in the universe beyond Earth, is one of the big open questions of our time. In the context of stress and adaptation to extreme conditions, it is of particular interest whether life might exist in places generally more hostile than our biosphere, such as our next-door neighbor planet Mars. In the context of the discussion of life's origins addressed in the previous chapter, we must ask whether life could have originated at different places in the universe simultaneously or whether life originated only once and spread through space and time. But first of all, we will face an utterly practical problem: Provided the aliens don't turn up to discover us, how can we detect extraterrestrial life?

How to Detect Life on a Planet

In December 1990, the NASA space probe *Galileo* passed a middle-sized planet at a minimal distance of just under 1,000 kilometers. Spectroscopic measurements performed by the instruments on board revealed a number of chemical abnormalities, most strikingly the presence of large amounts of molecular oxygen gas in the atmosphere, which contained traces of methane at the same time. This was clearly a state removed from chemical equilibrium, as the methane should react quickly with the oxygen to form the more stable compounds carbon dioxide and water. Furthermore, the absence of meteorite craters indicated continuous geological changes in the surface of the planet, which should expose oxidizable materials to the atmosphere and thus remove the oxygen. Parts of the surface had a remarkable capacity for absorbing red light. Finally, the probe also caught some irregular electromagnetic waves in the radio-frequency band emanating from the planet, which could be used by intelligent life forms for long-distance communication.

All these results, combined with the evidence for large amounts of liquid water covering more than half of the planet's surface, suggested that there is life on the planet in question. Now I guess it is time to admit that the group led by the astrophysicist Carl Sagan (see the Profile) that carried out these measurements with *Galileo* was cheating a little bit. The scientists knew right from the start that there was life on the planet they studied. It is the third planet of an ordinary yellow dwarf star of the spectral type G1 in the galaxy known as the Milky Way, and it happens to be the planet we live on and call Earth. Of course, the space probe had not been sent out to investigate whether there was life on Earth—for this, terrestrial methods would have been more appropriate, if one hadn't known the answer *a priori*, anyway. *Galileo* only used the Earth's gravitational field for a maneuver called "swing-by" in order to get catapulted swiftly to its target, Jupiter. The search for life on Earth was not planned before the launch of the probe, but improvised in the short term by Sagan and his co-workers.

What this unusual piece of research spiced with subtle irony showed quite impressively is the enormous difficulty of detecting life from a distance of just 1,000 kilometers (that is, from a very close encounter by astronomical standards) even on a planet that has been teeming with life for more than three billion years. There was no way the probe could have obtained visual evidence of buildings indicating the presence of advanced civilization. Carl Sagan and D. Wallace estimated that one would have to

 profile _____

Carl Sagan and the Quest for Life in the Universe

Without being an artist, he created one of the best-known graphical works of our century, the line drawing of a man and a woman combined with the position of our planet in the solar system, designed to be read as a calling card of our civilization by any aliens who might find it. Carl Sagan died in December 1996 at the age of 62, at the end of the very year in which the discovery of what may or may not be microfossils in a Martian meteorite had put the search for life in the universe, around which his life as a researcher orbited, into the spotlight of public interest. The very year, too, the investigation of other planets entered a new era with the launch of a major program of Mars voyages by NASA. Much of today's knowledge and excitement about the possibility of life in the universe can be safely called his personal achievement, but, sadly, he did not live long enough to get an answer to the question that inspired him for a lifetime.

Born in New York as a son of Russian immigrants, Sagan studied physics, taught at Harvard, and founded his own laboratory for planetary studies at Cornell in 1968. His countless contributions to our understanding of our own and other planets include the concepts of the greenhouse effect and of the nuclear winter, and many of them played a crucial role in NASA's space missions. He left his mark in all of the research fields discussed in this chapter. Furthermore, he will be remembered as one of the most productive and successful popularizers of science in our time, with an output of more than 20 books (including the Pulitzer prize-winning *The Dragons of Eden: Speculations on the Evolution of Human Intelligence* and the novel *Contact*, which became a successful movie in 1997) and numerous television programs. Sagan made it clear not only to scientists, but also to a wide audience that we are living on just a "pale blue dot" in the mind-blowing immensity of the universe.

photograph the entire surface of a planet with a resolution of two kilometers to detect a civilization like ours. *Galileo* could only obtain images of 2.3 percent of the Earth's surface at a resolution of one kilometer, and a further 4 percent with two kilometers resolution. No traces of civilization were found. Indeed, the radiowaves were the only sign of intelligent life that *Galileo* found.

How much more difficult will it then be to detect life on other, as-yet-undiscovered planets that may be lightyears away and/or only contain small biotopes well hidden away from cosmic radiation under the surface? The situation is not quite so desperate for our own solar system. One of its planets has time and again figured in discussions of potential extraterrestrial life—the fourth planet from the Sun, Mars.

Is There Life on Mars?

The fourth planet from the Sun (Figure 1) only tips the scales at one-tenth of the Earth's mass, but of all planets in our solar system, its climate is most similar to ours. A day on Mars (a "sol" for Mars enthusiasts) is just half an hour longer than over here. Temperatures near the equator can rise to a comfortable 24°C during the day, but beware of the night frost, which can be all of −80°C. This unfamiliarly large difference between day and night temperatures is partly due to the lack of oceans, which could serve as heat exchangers, and partly to the extremely thin atmosphere, which only amounts to half a percent of the terrestrial atmospheric pressure and thus cannot retain very much warmth either. At the poles, ice caps form in winter, containing water ice along with solid carbon dioxide.

Speculations about the possibility of life on Mars first arose when the Italian astronomer Giovanni Schiaparelli (1835–1910) discovered the long trenches that he called "*canali*" in 1877. Although he did not mean to imply that these structures were produced by intelligent beings, others translated *canali* as canals and interpreted them as drainage systems of a civilization trying to get water from the pole caps transported to the arid equatorial regions. Later, however, the Martian canals turned out to be just canyons, if not products of the imagination. Nevertheless, Martians figured regularly in the cast of science fiction literature and films, be it as invaders trying to colonize Earth, as in H. G. Wells' classic *War of the Worlds* (1898), or as an extinct civilization that left its monuments behind,

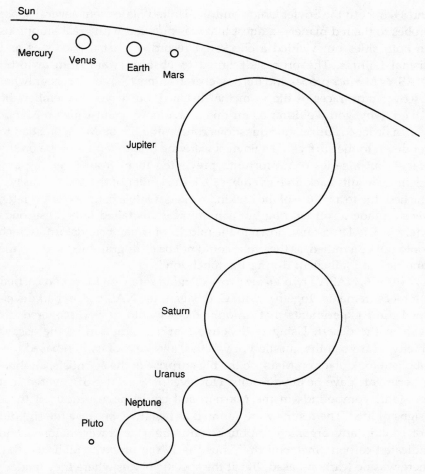

Figure 1. The planets of our solar system drawn to scale with the Sun, whose diameter is 1.5 million kilometers.

as in Ray Bradbury's *Martian Chronicles* (1950). And every self-respecting space-traveling future civilization of our planet entertains at least a gas station on Mars (which makes sense, because with the tenfold smaller gravitation, it's so much easier to start your space travel from there).

However, it took more than half a century before science could start to catch up with the imagination and travel to Mars in reality. Between 1962 and 1975, both the Soviet Union and the United States sent several space probes to the red planet—a quest that was pursued with great ambitions on both sides but yielded a checkered record of partial successes and dismal failures. The probes designed to observe Mars from an orbit (NASA's *Mariner* probes and the Soviet crafts *Mars 1–4*) could certainly not discover any traces of life from the distance, but if you remember the *Galileo* flyby, you wouldn't count this as evidence against life on Mars.

In order to address the questions concerning life on Mars, one had to get down to the surface. The Soviets were the first to land a probe on the planet, but a series of misfortunes prevented them from receiving any scientific results back. In November 1971, the lander of the *Mars 2* mission crashed due to a failure of the braking rockets. Only a couple of days later, *Mars 3* made a soft landing but communications failed only 20 seconds afterwards. *Mars 5* came down in the middle of a major sandstorm (which could not be avoided, as the probe could not be reprogrammed to defer the landing) and fell silent during the touchdown.

In 1976, NASA's two *Viking* probes landed safely on Mars and carried out measurements. Together with Carl Sagan, the NASA team had developed three experiments that on our planet would have discovered life even in the desert. Using radioactive marker compounds, the assays chiefly addressed the question of whether any kind of carbon-based metabolism took place on Mars. To the big surprise of the scientists, all three experiments gave positive results. However, it was later discovered that inorganic compounds in the Martian soil could have produced these "signs of life." The results were counted as negative, because the soil did not contain any organic compounds, although the tracer methods had indicated carbon "metabolism." Thus the *Viking* probes established that there was no (carbon-based) life at the specific places where they landed, but this finding does not exclude that life exists or has existed elsewhere on Mars. The Viking probes continued to send data to Earth for another five years, so their investigation of Martian soil and atmospheric composition can be counted as a major success (even though they are best remembered for their failure to find life on Mars).

However, the nonevidence for life and the general impression that Mars was too hostile to bear life at all discouraged further investigations, and there was no American traffic from Earth to Mars between 1976 and 1993. When the program was resumed, the search for life on the red planet suffered a further drawback when the probe *Mars Observer* was lost in

August 1993. Most probably, the probe exploded while attempting to enter orbit around Mars.

The tide began to turn in the fall of 1995, when the discovery of autotrophic microorganisms (which are independent of organic nutrients) in basalt layers deep under the surface of the Earth suggested that such "stone-eaters" might also exist on Mars (see Chapter 2). In the planet's youth, some 3.8 to 4 billion years ago, conditions may have been quite similar on both planets and life could have originated in parallel or spread from one to the other. When things got worse on Mars—partly due to the fact that the smaller planet lost most of its gases to space and thus could provide little protection from radiation and temperature changes—autotrophic microbes may have sought shelter beneath the surface. At a reasonable depth, they would have experienced quite stable temperatures, possibly even with a geological heat supply.

These considerations, along with the accumulating knowledge about terrestrial extremophiles, have led many scientists to change their minds concerning life on other planets. Twenty years after the disappointment of the *Viking* missions, no one would want to exclude the possibility of past or present life on Mars. In January 1996, for instance, the Ciba Foundation in London hosted a meeting on the origins of life on Earth and—possibly— on Mars. NASA initiated a new exploration program that began with the launch of two probes, *Mars Global Surveyor* and *Mars Pathfinder*, in November and December 1996, respectively. The latter landed safely on Mars on 4 July 1997 and started to run a solar-powered microlaboratory with a minirover collecting samples, while the former will orbit the planet to survey its surface with a resolution of one meter (see Sidelines, "First Results of *Mars Pathfinder* and *Mars Global Surveyor*," for a last-minute report on the findings of the two probes). Further probes will be launched at 26-month intervals, that is, each time the most favorable constellation of the two planets returns. The final probe of the program is due to bring back Martian soil in 2005.

However, it may not be necessary to fly to Mars in order to investigate Martian minerals. Every once in a while, small lumps of stone catapulted from Mars into space by a meteorite impact or volcanic eruption come down on Earth. A group of meteorites called the SNC meteorites (after the initials of the places where the first three of them were found) originated some 1.1 to 1.3 billion years ago by volcanic activity and received a space lift from meteorite impacts some time later. Twelve such meteorites have

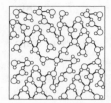

sidelines _____

First Results of *Mars Pathfinder* and *Mars Global Surveyor*

No, they have not discovered life on Mars, nor are they likely to do so in the months to come. But then, neither of the two probes that successfully started the new NASA Mars program was meant to look for life. Their target was to demonstrate that the red planet is accessible to modern science.

The official mission objectives proclaimed by NASA for the *Mars Pathfinder* probe included the demonstration that space exploration can be "faster, better and cheaper" than previously thought. With just three years of development time and a price comparable to a blockbuster movie, this demonstration was a major success from day one, when the probe came bouncing down to the surface of our favorite planetary neighbor on 4 July 1997. The real-life action of its dramatic touchdown and first explorations on Mars attracted millions of viewers to television screens worldwide, and during the following weeks the *Pathfinder* website with its dozens of mirror sites all over the world became the busiest place in the world-wide web.

As a bonus, the science systems of the probe actually worked and collected heaps of chemical, geological, and meteorological data. While this book is going into production (September 1997), the cute solar-powered toy car known to the world as *Sojourner* is about to retire after more than two months spent wandering around from rock to rock and analyzing the chemical composition of the Martian minerals using a tool known as an alpha proton X-ray spectrometer. Basically, it contains a tiny source of radioactivity that bombards the sample in question with alpha particles (two protons plus two neutrons, corresponding to the nucleus of a helium atom), which triggers different kinds of response in different chemical elements.

The detailed analysis of the data coming in from Mars will keep scientists busy for years, but a few trends can already be distinguished. The overall composition of Martian soil seems to be very similar at the landing sites of *Pathfinder* and *Viking*, which are more than 1,000 kilometers apart. The new data on its elemental composition are also in agreement with the analysis of the SNC meteorites (which were identified as of Martian origin by comparison with *Viking* data). However, the composition of one of the objects that *Sojourner* investigated, a small rock named Barnacle Bill, turned out to be more Earth-like than any other Martian mineral known so far. This may be due to a history of heating in the presence of water—which, if confirmed, would be good news for the possibility of life on early Mars.

Apart from rocks with fancy names and the dust that they classified into two categories, *Pathfinder* scientists are mostly talking about the weather. Forecasts must be even more difficult on Mars than over here, as the temperature and pressure of the thin atmosphere can vary spectacularly over short distances and time intervals. Using instruments mounted on its mast, *Pathfinder* found that the nose and toes of a human visitor might experience temperature differences of up to 40°C. And the values can change by up to 20°C within a couple of minutes. Similarly, the atmospheric pressure and the winds are more variable than on our home planet. A systematic understanding of the weather on Mars will, however, be crucial for future missions, which will include long-distance travel across the planet by balloon.

The maps for these expeditions will be drawn according to the data that *Mars Global Surveyor* will send back to Earth from its orbit. Even before it reached Mars as scheduled on 11 September 1997, the probe was already useful, taking pictures of the planet from afar prior to *Pathfinder*'s landing. By the time this book is printed, more results from both probes will have hit the headlines. Their overwhelming success certainly sets the gears of the Mars exploration program to an exciting future, which will also include a reevaluation of the questions concerning past life on our neighbor planet.

---

been proven to come from Mars. This has been possible because the *Viking* probes measured details of the isotope characteristics of Martian soil and atmosphere. While some of the SNC meteorites even contained gas inclusions that could be identified as Martian "air," all of them matched the isotope content of Martian soil.

One of these meteorites, known as ALH84001, stirred a major media hype when NASA scientists announced they had found fossils resembling those of terrestrial bacteria in that lump. *Science* magazine waived its usual embargo on prepublication publicity some 10 days before the scheduled publication date of 16 August 1996 after rumors about the forthcoming paper had appeared in the press. NASA made best use of this chance and even got President Clinton to announce the good news—and the rest is history[1]. It was one of the very rare occasions that a scientific discovery made it to the front pages of almost every newspaper and magazine.

The plain, fist-sized stone that made so many waves had been discovered on the Allen–Hills ice field in Antarctica in 1984 (hence the name ALH84001), but its Martian origin was only recognized in 1993. It is thought to have solidified some 4.5 billion years ago; thus it is much older than the other SNC meteorites. Its space voyage began about 16 million years ago and ended ca. 11,000 BC in the ice of the antarctic. Mineralogical investigations and the determination of the isotopic distribution of the carbon contained in the stone have established that it has not suffered weathering to any significant extent since it came down.

These are the findings that made NASA scientists think it contained microfossils:

- At the interior fracture surfaces of the metorite, complex organic molecules, so-called polycyclic aromatic hydrocarbons (PAH) were found. Both the concentration and the molecular composition of the PAHs exclude the possibility of terrestrial contamination.
- In contrast to other SNC meteorites, ALH84001 contains carbonate globules with diameters varying from 1 to 250 micrometers. They are estimated to be some 3.5 billion years old, i.e., considerably younger than the surrounding rock. Some of them have suffered shock deformations, which can only have happened on Mars or in space, not in Antarctica.
- In the vicinity of the globules, both oval and irregularly shaped structures were found that resembled terrestrial microfossils (Figure 2). With a maximum length of 100 nanometers, however, they are smaller than all known microorganisms.
- Furthermore, the meteorite contains small-grained iron sulfide and magnetite particles resembling those made by magnetotactic bacteria.

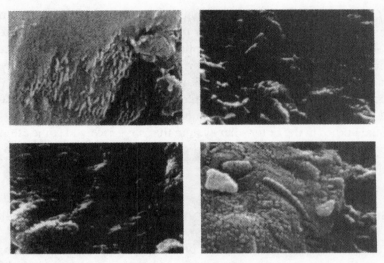

Figure 2. Electron micrographs of the fracture surfaces of the Martian meteorite ALH84001. Close to the globular carbonate deposits, there are fibrillar structures resembling terrestrial microfossils. However, with lengths up to 100 nanometers, they are smaller than known bacteria.

The authors of the *Science* report admit that none of these pieces of evidence is inevitably of biological origin. However, as they conclude, "Although there are alternative explanations for each of these phenomena taken individually, when they are considered collectively, particularly in view of their spatial association, we conclude that they are evidence for primitive life on early Mars."

This conclusion has sparked some controversy, but the limited amount of material to draw evidence from may mean that the question cannot be settled decisively at this point. Results from the current program of Mars probes may help to explain the origin—biological or not—of the remarkable features found in this meteorite.

Considering the possibility that life on Mars, if it ever existed, presumably never surpassed the evolutionary stage of unicellular organisms and may have died out three billion years ago, the search for its traces will be challenging even with the best instruments. In any case, planetary scientists have learned from the extremophilic microbiologists that one should never say never.

Conditions on all the other planets of our solar system are even more hostile than on Mars. The two inner planets, Mercury and Venus, are too hot. Although Venus has an atmosphere of carbon dioxide and traces of water vapor, temperatures at its surface can reach 260°C. And beyond Mars, the energy supply from the Sun becomes too weak to sustain life. Apart from that, the gas giants Jupiter and Saturn don't even possess a solid surface where organisms could settle. However, there are other celestial bodies in the outer solar system that have received some attention in the origin-of-life discussion, particularly Jupiter's moon Europa and Saturn's Titan.

Strange Worlds: The Moons of the Big Gas Planets

At an average distance of 1.5 billion kilometers from the Sun, the ringed planet Saturn is the outmost of the planets known to mankind from ancient times. Among its numerous satellites, Titan is the "oldest" in that it was discovered as early as 1655, the largest—with its 5,000-kilometer diameter it is in fact slightly bigger than the planet Mercury, and the only one to carry an atmosphere worth mentioning. Like ours, it contains mainly molecular nitrogen, but it is 10 times more dense. Like the primeval atmosphere of our planet, it is reducing, with methane as the second most abundant constituent. It also contains various simple organic molecules such as hydrocarbons and nitriles. Large amounts of water may be hidden as ice beneath the surface of Titan.

Carl Sagan and Reid Thompson at Cornell demonstrated that radiation such as that to which Titan's atmosphere is exposed can indeed stimulate the formation of organic molecules from nitrogen and methane. Other investigations from Sagan's laboratory showed that further radiation-induced reactions in Titan's atmosphere can lead to the synthesis of a solid that Sagan called Titan-tholin. This substance may in fact be the cause of the reddish fog that barred the view to the moon's surface when *Voyager 2* passed it in 1981. If the reactions simulated by Sagan and the formation of Titan-tholin have been going on for four billion years, the surface of Saturn's satellite must be covered with a layer of organic substance possibly hundreds of meters thick.

Sagan also suggested that due to the energy intake from meteorite impacts, every spot on Titan must have been covered by liquid water at some time in its history. In the laboratory, Titan-tholin and water can react to form amino acids, nucleotide bases, and PAHs—suggesting that the moon's surface harbors an ideal set of molecules for the origin of life, and one not so very different from the results of Miller's primeval soup experiment (Chapter 5). Could the local heating caused by meteorite impacts have persisted for long enough to allow the emergence of life from those simple prebiotic building blocks? For the answer, we will have to wait a little while. The next visit to Titan is scheduled for November 2004. As part of the *Cassini* mission, a project shared by the American, European, and Italian space agencies, the probe *Huygens* will land on Titan.

Similarly, Jupiter's satellite Europa is considered a hot candidate for extraterrestrial life. Images obtained from a close flyby of the *Galileo* probe in January 1997 revealed that the icy surface of the moon seems to be in permanent upheaval, which may be due to heat-induced convection. If there is sufficient fire under the ice, liquid fresh water may be hidden underneath the ice shield, and life might exist or have existed there. In a way, the potential lakes underneath the icy surface of Europa may resemble those found in Antarctica (see Chapter 2). If life was discovered underneath the antarctic ice shield, Europa would become the most promising candidate as a site for extraterrestrial life in our solar system.

Are There Any Planets Orbiting Other Stars?

Now we have covered all the candidates for potential extraterrestrial biotopes in our solar system. If you want to discover a twin of our planet with highly diverse and developed life, you will have to look further afield, in other solar systems. The only trouble is, no one has ever seen a planet orbiting a star that is not our Sun. Until 1995, there weren't even indirect hints of other planetary systems. When scientists discussed the possibility of such systems, their only argument in favor of their existence was a probability consideration. It would have been really strange if the universe with its millions of stars would have given rise to one and only one planetary system. According to common theories about the origin of the solar system, the planets aggregated from a flat disk of dust and gas particles surrounding the young Sun and reaching far out even beyond the orbit of Pluto. Exactly the same kind of disks have been observed around countless young stars. If their development normally leads to the formation of planets, there should be a huge number of stars with planetary systems.

But where are all these planets hiding? As a matter of principle, it is extremely difficult to discover a relatively small, nonradiant body in the vicinity of a big, shining star, while the aforementioned gas disks are much more convenient to find—at certain wavelengths they are even brighter than their stars. Even for stars in our direct neighborhood, one cannot be sure that a planet of the size of Jupiter would be directly observable. However, small deviations in the star's movement induced by the orbiting planet should be detectable. With the help of the wavelength shift of moving objects (the acoustic equivalent of which is well known from the change in pitch of a passing ambulance, for instance [Figure 3]) this gravitational pull can be observed. This requires, however, that the planet is quite big and orbits its star in a relatively short time, so that measurements of the star's Doppler shift can be made over a whole orbital period with sufficient precision.

This consideration was well known but remained pure theory until in October 1995 Michel Mayor and Didier Queloz of the Geneva Observatory announced that they had discovered a giant planet, half the size of Jupiter, orbiting the Sun-like star 51 Pegasi at an extremely short distance in just 4.2 days. As the orbit of this planet only measures one-eighth of that of

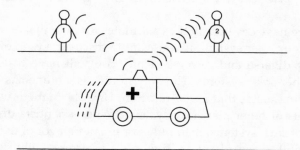

Figure 3. The Doppler effect, which astronomers use to detect relative motion of stars by changes in the wavelength of their light, can be more readily observed with sound waves. The siren of an ambulance, for instance, will be heard at a higher pitch (shorter wavelength) while it is approaching (person 2), and at a lower pitch (longer wavelength) after it has passed (person 1). In optical Doppler shifts, longer wavelengths mean a move to the red end of the spectrum, shorter wavelengths a shift toward blue.

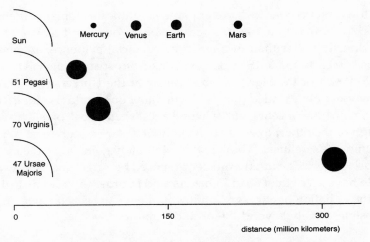

Figure 4. Schematic overview of the first extrasolar planets to be discovered by indirect methods. The sizes of the bodies are not to scale with the distances between them.

Mercury, the surface must be white hot (1,300°C), far too hot even for extremophiles in any case (Figure 4).

Shortly afterward, other astronomers confirmed the evidence for a planet orbiting 51 Pegasi and discovered two more stars with suspicious Doppler shifts. One of the presumed planets, twice the weight of Jupiter, is in an orbit around the star 47 Ursae Majoris that would be equivalent to a position in our solar system between Mars and Jupiter. The other one is even six and a half times heavier than Jupiter and circles around 70 Virginis within the "Goldilocks zone," where liquid water could exist at its surface. However, astronomers are uncertain about whether this body is in fact a planet or rather a brown dwarf, that is, a gas ball that originates like a star but does not have the critical mass to shine.

These discoveries triggered a kind of gold rush in the observatories, and new discoveries were made faster than the print journals could report them. The Paris Observatory launched an automatic catalogue of new planets on the world-wide web, the "Extrasolar Planets Encyclopedia." By the end of September 1996, the list contained 18 suspicious bodies; by the end of May 1997 there were 22 confirmed planets (possibly including some brown dwarfs) and a further 10 doubtful objects.

However, in order to discover planets significantly smaller than Jupiter and/or with orbital periods comparable to ours, and maybe observe them directly rather than through circumstantial evidence, astronomers will be forced to make their observations from space. Roger Angel and Neville Woolf of the Steward Observatory at the University of Arizona have worked out detailed plans showing how one could use interference phenomena (the canceling out of light waves when a peak of one wave hits a trough of the other) to weaken the light of a star enough to stop it from outshining its satellites. "Think big!" is definitely going to be the motto of any such project. The interferometer proposed by Angel and Woolf would have to have a length of 50 to 75 meters, and in order to avoid disturbance from the radiation of our own Sun it would have to be as far away from it as possible, ideally beyond the orbit of Jupiter.

The Spore's Guide to the Galaxy

As yet, nobody knows whether there is extraterrestrial life in the universe, but the strongest argument of those who believe in it is the relatively young age of our solar system. As the universe is at least four times older, biospheres and civilizations may have arisen and perished in other solar systems long before ours even existed. And this earlier life could well have spread out, and some people even think it seeded life on our planet. In the context of research on the origin of life, this so-called panspermia hypothesis is notoriously counterproductive. By relocating the event of the origin of life some billions of years back in time and some lightyears away in space, it removes any chance of finding out how it happened.

However, a little speculation about the ways in which life could have traveled through time and space may be quite amusing. In 1996, J. Secker, J. R. Lepock, and P. S. Wesson, at McMaster University, Hamilton, and Waterloo University, Waterloo (both Ontario, Canada), managed to put this speculation onto a reasonably solid foundation. They carried out calculations concerning the conditions under which microbial spores could be propelled through space by the radiation pressure of stars. The main problem is that the life-threatening ultraviolet radiation would kill off any germ floating through space unprotected in a matter of days. Secker and co-workers therefore dressed their hypothetical traveler in a coat of carbon dust, a material commonly found in certain meteorites. This already requires a cost–benefit evaluation comparable to the task of building a spaceship. If the graphite coat gets too thick, the coated germ will get too heavy for the radiation pressure to accelerate it to the critical escape speed required to get out of the gravitational field of the home star. The best launch pad for such a voyage is a planet orbiting a red giant. These stars shine up to 100 times brighter than our Sun, and their radiation is mostly at longer wavelengths with relatively little contribution of the dangerous UV radiation. Furthermore, they are older than our Sun, which will need another five to ten billion years before its becomes a red giant itself, so there would have been enough time for the evolution of life on their planets.

Once the traveling microbes leave the realm of their home star and drift through space for a few million years, they may be lucky enough to encounter another solar system with planets. If many such particles rain down onto a planet like ours, there would be a realistic chance that some of them would make their landing at just the right angle and speed to crack the graphite coat without killing the passenger (Figure 5). Secker and co-

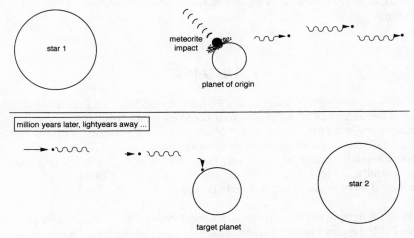

Figure 5. The panspermia hypothesis in the variant discussed here, where the radiation pressure of a star effects the transport of germs through interstellar space. When the germs arrive in a different solar system, the radiation pressure of the target star will slow down their movement and—with a bit of luck—they can land softly on a planet.

workers have estimated that microorganisms can thus travel a distance of some 20 lightyears in a million years. Longer travel times would bring the survival chances down to zero, so it is recommended to touch down on a hospitable planet every once in a while. Once the germ has spread over most of the new planet, its descendants could take off with the next meteorite impact or volcanic eruption to continue the star trek. This way, life could have spread across most of the Milky Way by now—if there ever was extraterrestrial life in the Milky Way, and a reasonable number of habitable planets, that is. That "habitable" does not necessarily mean 20°C, 1 bar atmospheric pressure, and 20 percent oxygen in the atmosphere is the lesson astronomers have learned from biologists in recent years. And the biologists learned it from the extremophiles.

Endnote

1. For many observers, the really interesting question was not the authenticity of the fossils, but the *cui bono* of the media hype. Potential beneficiaries included President Clinton, striving for reelection, NASA, hoping for taxpayers' cash for its ambitious Mars program, and last but not least the producers of the blockbuster *Independence Day*, also known as the most expensive B-movie of all times.

Updates

p. 161   Following the successful 1997 arrivals of Mars Pathfinder and Mars Global Surveyor, NASA suffered a series of Mars-related disasters in 1999, culminating in the loss of the probe *Mars Climate Orbiter* after a series of errors starting from a failure to convert imperial measures to metric ones. In early 2000, the discussion about the possibility of life on Mars was revived by the suggestion that water has quite recently shaped its surface, and that the red planet might contain somewhat more water than previously thought.

p. 163   A flyby of the NASA probe *Galileo* in January, 2000 recorded a frequent reversal of the moon's magnetic field. This could best be explained by the presence of a conductive liquid, such as salt water, underneath the frozen crust.

7

Glossary

Acidophilic Acid-loving.

Adjuvant An empirically optimized mixture of chemicals and cell extracts enhancing the immune response.

Aerobic Able to grow in the presence of oxygen. See anaerobic.

Alkalophilic Base-loving.

Anaerobic Only able to survive in the strict absence of oxygen.

Antigen A molecule or part of a molecule that triggers an immune response following molecular recognition by an antibody.

Aquifer Soil layer conducting groundwater.

Archaebacteria (archaea) One of the three domains of life, along with the eukaryotes (eukarya) and eubacteria (bacteria).

Bacteriophage (phage) A virus which infects bacteria. Phage genetics is one of the oldest and best-studied areas of molecular biology.

Barophilic Pressure-loving.

Basalt   Generic name for a group of usually dark, fine-grained lava minerals occurring in countless variants. The major constituents are oxides of the elements silicon, titanium, aluminum, and iron.

Bioluminescence   Generation of light from metabolic energy. Although fireflies are the best-known example, bioluminescence is also found in dozens of other species including fish, jellyfish, and mushrooms.

Biosphere   Those parts of the Earth where organisms can live; the biologically habitable parts of soil, air, and water.

Black smoker   Nickname of → hydrothermal vents.

Catalysis   Acceleration of a chemical reaction by lowering the energy barrier. The strict definition of catalysis includes that the catalyst must not be affected by the overall reaction.

Crust   The outermost shell of the Earth, reaching 25 to 40 kilometers deep.

Cytoplasm   The liquid phase in the interior of a cell (excluding → organelles).

DNA   Biological chain molecule built from four kinds of building blocks, the nucleotides. DNA, normally found as a double helix formed by two complementary strands, is the carrier of genetic information in all cellular life forms and many viruses.

DNA replication   Copying of DNA by specialized enzymes, the DNA polymerases. For this purpose, the double helix structure is dissolved, and a new complementary strand is synthesized along each strand.

DNA transcription   Synthesis of → RNA following the genetic information encoded in DNA. In contrast to → DNA replication, the transcription only makes use of one of the two DNA strands, the sense strand.

Enzyme   A protein that can perform → catalysis, i.e., speed up a chemical reaction without getting changed itself.

Eubacteria   See archaebacteria.

Eukaryotes   See archaebacteria.

Expression   Realization of the genetic information encoded in DNA leading to the synthesis of a protein or a stable RNA. The word is used in particular if a foreign gene is introduced into a host cell which then expresses the gene, i.e., produces a recombinant protein, or if the rate of synthesis is artificially enhanced (overexpression).

Gaia theory   A hypothesis first proposed by James Lovelock, which states that the biosphere of our planet is a self-regulating cybernetic system, which can be understood as a hyperorganism. Lovelock named this system after the ancient Greek earth goddess Gaia.

Genetic code   The (almost) universally valid rules by which three-letter "words" in → DNA or → RNA are assigned to amino acids to be incorporated into → proteins.

Genome   All the genes of a particular organism.

Geyser   A hot spring that does not flow continuously, but discharges at regular intervals.

Halophilic   Salt-loving.

Heat shock proteins   A group of proteins that are synthesized in large amounts after a short overheating of a cell. Some of them have important functions in the context of protein folding (→ molecular chaperone).

Helix   Structure wound up like a screw. In biochemistry, helices are commonly found in proteins (alpha-helix) and in nucleic acids (double helix).

Hydrophilic   Water-loving, i.e., well soluble in water (of molecules and parts of molecules).

Hydrophobic   Water-avoiding.

Hydrophobic interaction   Tendency of hydrophobic molecules or parts of molecules to cluster with other hydrophobic groups to minimize their exposure to water. Hydrophobic interactions are important factors for the stability of cellular structures such as the double-layer membrane and the inner core of proteins.

Hydrothermal vents   Hot springs occurring in volcanic regions of the ocean floor. The heavy-metal ions and hydrogen sulfide dissolved in the overheated vent fluid precipitate as metal sulfides as soon as the contact with seawater cools the fluid. This reaction produces the characteristic black "smoke" plume. The vent chimney also builds up from precipitated materials, mostly gypsum and sulfides.

Hyperthermophilic   Extremely heat-loving, normally used to describe microorganisms whose optimal growth temperature lies above 80°C.

Ion channels   Pores in the cell membrane that allow and regulate the transport of ions across the membrane.

Macromolecules   Relatively big molecules containing hundreds to ten thousands of atoms. Typically → polymers of smaller building blocks.

Magnetotactic bacteria   Bacteria that can use the Earth's magnetic field for orientation. They contain minute crystals of magnetic iron minerals enclosed in a specialized cell compartment.

Mesophilic   Opposite of extremophilic, i.e., "normal."

Messenger RNA (mRNA)   The information carrier used by the → ribosome during → protein biosynthesis. The mRNA arises from the process of → DNA transcription. In contrast to → transfer RNA and ribosomal RNA, it is quite short-lived. The latter two, therefore, may be called stable RNAs.

Methanogenic bacteria   A group of organisms in the domain of → archaebacteria, which convert carbon dioxide and hydrogen to methane and water.

Molecular chaperone   A protein which helps unfolded or freshly synthesized proteins to fold to the correct three-dimensional structure by suppressing unwanted side reactions.

Organelles   Compartments of the cell of → eukaryotes, which are separated from the → cytoplasm by a membrane and fulfill specialized functions. For example, mitochondria (energy metabolism), chloroplasts (photosynthesis).

Osmosis   Passage of dissolved molecules or ions through a porous wall (membrane) between two liquids. In particular, the tendency to equilibrate concentrations of dissolved substances on either side. If the membrane is permeable only for the solvent, the latter flows to the side of the higher substance concentration and builds up an osmotic pressure.

Oxidation   A reaction in which the molecular group considered gives electrons away (i.e., its oxidation number increases). The name is due to the fact that the first examples of such reactions, such as the combustion of carbon compounds to carbon dioxide or the rusting of iron, involved oxygen. The reaction partner which accepts the electron is reduced, i.e., its oxidation number is decreased. The whole process can also be called a redox reaction.

Photochromic material   A substance that changes color upon illumination (as in light-adapting sunglasses).

Polymer   A molecule built from a large number of similar building blocks (monomers). Polymers which can be described as a regular repetition of one or a few monomers are called homopolymers. These include common plastics such as polystyrene, polyethylene, etc. In contrast, biological polymers ($\rightarrow$ DNA, $\rightarrow$ proteins) tend to consist of an irregular (information-containing) sequence of similar building blocks and are therefore labeled heteropolymers.

Polymerase chain reaction (PCR)   A method for copying $\rightarrow$ DNA, which can be applied to very small samples, provided the sequences flanking the gene of interest are known or can be guessed. By repetitive application of the PCR cycle, an exponentially growing number of copies of the original DNA can be made.

Protein biosynthesis   The process of linking up amino acid building blocks to form a $\rightarrow$ protein molecule. It is carried out by the $\rightarrow$ ribosome together with a set of other cellular factors.

Protein folding   The process in which the linear chain of amino acids (also called a polypeptide) arising from $\rightarrow$ protein biosynthesis or from denaturation of a protein forms a three-dimensional structure stabilized by many weak interactions. Most proteins have to be folded to be biologically active.

Proteins   Chain molecules consisting of (dozens to thousands of) amino acid building blocks and carrying out a wide variety of tasks in the cell, including catalysis of metabolic reactions, transport of small molecules or ions, mechanical work, switch functions, and information transfer.

Psychrophilic   Cold-loving.

Ribosome   A complex of more than 50 $\rightarrow$ proteins and several $\rightarrow$ RNA molecules, which carries out the synthesis of proteins following the genetic instructions read from the $\rightarrow$ messenger RNA with the help of $\rightarrow$ transfer RNAs and various protein factors.

Ribozyme   An $\rightarrow$ RNA molecule that can perform $\rightarrow$ catalysis.

RNA   A biological chain molecule which acts as a mediator between genetic information ($\rightarrow$ DNA) and function ($\rightarrow$ proteins). The most important kinds of RNA are $\rightarrow$ messenger RNA (mRNA), $\rightarrow$ transfer RNA (tRNA), and the RNA of the $\rightarrow$ ribosome (rRNA).

RNA-world   A hypothetical stage in the early evolution of life, with RNA molecules as the only biological macromolecules playing the roles of both information carrier and function molecules.

Solfatara   Gas source in volcanic soil.

Sporulation   Formation of a special kind of cell, the inactive, but stress-resistant spore.

Thermophilic   Heat-loving.

Transfer RNA   Stable → RNA molecules which act as specific carriers for the amino acid molecules to be incorporated in → protein biosynthesis.

Transcription   Copying of the genetic information from → DNA onto → RNA by an enzyme called RNA polymerase. The regulation of transcription by specialized proteins, the transcription factors, is a central switchboard for all life processes in a cell.

Translation   Just a different word for → protein biosynthesis, used to emphasize the aspect that the ribosome "translates" the three-letter → genetic code used in → DNA and → RNA into the 20-letter amino acid code of the proteins.

Wild type   Organisms of a given species which carry the standard version of a certain gene as opposed to mutants, which carry an altered version of it.

8

Further Reading and Internet Links

Note to the reader: The author's home page address is http://www.
michaelgross.co.uk

Introduction: Life and Its Limits

Frausto da Silva, J. R. R., and Williams, R. J. P., *The Biological Chemistry of the Elements. The Inorganic Chemistry of Life* (Oxford: Oxford University Press, 1991).

Extreme Environments and Their Inhabitants

Thermophiles

Blöchl, E., Rachel, R., Burggraf, S., Hafenbradl, D., Jannasch, H. W., and Stetter, K. O., "*Pyrolobus fumarii*, gen. and sp. nov., represents a novel group of archaea, extending the upper temperature limit for life to 113°C," *Extremophiles* 1 (1997), 14–21.

Brock, T. D., "The road to Yellowstone—and beyond," *Annual Reviews in Microbiology* 49 (1995), 1–28.

Brock, T. D., and Madigan, M. T., *Biology of Microorganisms*, 7th ed. (Englewood Cliffs, New Jersey: Prentice-Hall, 1994).

Edmond, J. M., and von Damm, K., "Hot springs on the ocean floor," *Scientific American* 248 (April 1983), 70–85.

Huber, R., Kurr, M., Jannasch, H. W., and Stetter, K. O., "A novel group of abyssal methanogenic archaebacteria (*Methanopyrus*) growing at 100°C," *Nature* 342 (1989), 833–834.

Huber, R., Stoffers, P., Cheminee, J. L., Richnow, H. H., and Stetter, K. O., "Hyperthermophilic archaebacteria within the crater and open-sea plume of erupting Macdonald Seamount," *Nature* 345 (1990), 179–181.

Rees, D. C., and Adams, M. W. W., "Hyperthermophiles: Taking the heat and loving it," *Structure* 3 (1995), 251–254.

Shanks, W. C., "Mid-ocean ridges: Rebirth of a sea-floor vent," *Nature* 375 (1995), 18–19.

Trent, J. D., Chastain, R. A., and Yayanos, A. A., "Possible artifactual basis for apparent bacterial growth at 250°C," *Nature* 307 (1984), 737–740.

Tunicliffe, V., and Fowler, M. R., "Influence of sea-floor spreading on the global hydrothermal vent fauna," *Nature* 379 (1996), 531–533.

White, R. H., "Hydrolytic stability of biomolecules at high temperatures and its implication for life at 250°C," *Nature* 310 (1984), 430–432.

Psychrophiles

Ancel, A., Visser, H., Handrich, Y., Masman, D., and Le Maho, Y., "Energy saving in huddling penguins," *Nature* 385 (1997), 304–305.

Eastman, J. T., and De Vries, A. L., "Antarctic fishes," *Scientific American* 255 (November 1986), 96–103.

Ellis-Evans, J. C., and Wynn-Williams, D., "Antarctica: A great lake under the ice," *Nature* 381 (1996), 644–646.

Kapitsa, A. P., Ridley, J. K., Robin, G. d. Q., Siegert, M. J., and Zotikov, I. A., "A large deep freshwater lake beneath the ice of central East Antarctica," *Nature* 381 (1996), 684–686.

Radok, U., "The Antarctic ice," *Scientific American* 253 (August 1985), 82–89.

Thomas, D., and Dieckmann, G., "Life in a frozen lattice," *New Scientist* (11.6.1994), 33–37.

Life under Pressure

Cioni, P., and Strambini, G., "Pressure effects on protein flexibility: Monomeric proteins," *Journal of Molecular Biology* 242 (1994), 291–301.

Deming, J. W., "Ecological strategies of barophilic bacteria in the deep ocean," *Microbiological Sciences* 3 (1986), 205–211.

Deming, J. W., et al., "Isolation of an obligately barophilic bacterium and description of a new genus, *Colwellia gen. nov.*," *Systematic and Applied Microbiology* 10 (1988), 152–160.

Gross, M., and Jaenicke, R., "Proteins under pressure: The influence of high hydrostatic pressure on structure, function and assembly of proteins and protein complexes," *European Journal of Biochemistry* 221 (1994), 617–630.

Jannasch, H. W., and Taylor, C. D., "Deep-sea microbiology," *Annual Reviews in Microbiology* 38 (1984), 487–514.

Somero, G. N., "Biochemical ecology of deep-sea animals," *Experientia* 48 (1992), 537–543.

Yayanos, A. A., "Microbiology to 10,500 meters in the deep sea," *Annual Reviews in Microbiology* 49 (1995), 777–805.

Yayanos, A. A., and Dietz, A. S., "Death of a hadal deep-sea bacterium after decompression," *Science* 220 (1983), 497–498.

Yayanos, A. A., Dietz, A. S., and van Boxtel, R., "Obligately barophilic bacterium from the Mariana Trench," *Proceedings of the National Academy of Sciences of the USA* 78 (1981), 5212–5215.

Bioluminescence

Hastings, J. W., "Biological diversity, chemical mechanisms, and the evolutionary origins of bioluminescent systems," *Journal of Molecular Evolution* 19 (1983), 309–321.

McElroy, W. D., and Glass, B., *Light and Life* (Baltimore: The Johns Hopkins Press, 1961).

Deep Subsurface

Fliermans, C. B., and Balkwill, D. L., "Microbial life in deep terrestrial subsurfaces," *BioScience* 39 (1989), 370–377.

Fyfe, W. S., "The biosphere is going deep," *Science* 273 (1996), 448.

Kaiser, J., "Microbiology: Can deep bacteria live on nothing but rocks and water?" *Science* 270 (1995), 377.

Sarbu, S. M., Kane, T. C., and Kinkle, B. K., "A chemoautotrophically based cave ecosystem," *Science* 272 (1996), 1953–1955.
Stevens, T. O., and McKinley, J. P., "Lithoautotrophic microbial ecosystems in deep basalt aquifers," *Science* 270 (1995), 450–454.

Drought Resistance

Mattimore, V., and Battista, J. R., "Radioresistance of *Deinococcus radiodurans*: Functions necessary to survive ionizing radiation are also necessary to survive prolonged desiccation," *Journal of Bacteriology* 178 (1996), 633–637.
von Willert, D. J., "*Welwitschia mirabilis* Hook. fil.—das Überlebenswunder in der Namibwüste," *Naturwissenschaften* 81 (1994), 430–442.

Halophiles

Steinhorn, I., and Gat, J. R., "The Dead Sea," *Scientific American* 249 (October 1983), 102–109.

Acido- and Alkalophiles

Darland, G., Brock, T. D., Samsonoff, W., and Conti, S. F., "A thermophilic, acidophilic mycoplasma isolated from a coal refuse pile," *Science* 170 (1970), 1416–1418.
Hoffmann, A., and Dimroth, P., "The electrochemical proton potential of *Bacillus alcalophilus*," *European Journal of Biochemistry* 201 (1991), 467–473.
Krulwich, T. A., Hicks, D. B., Seto-Young, D., and Guffanti, A. A., "The bioenergetics of alkalophilic bacilli," *Critical Reviews in Microbiology* 16 (1988), 15–36.
Schleper, C., *et al.*, "*Picrophilus* gen. nov., fam. nov.: A novel aerobic, heterotrophic, thermoacidophilic genus and family comprising archaea capable of growth around pH 0," *Journal of Bacteriology* 177 (1995), 7050–7059.
Schleper, C., Pühler, G., Kühlmorgen, G., and Zillig, W., "Life at extremely low pH," *Nature* 375 (1995), 741–742.
Sletten, O., and Skinner, C. E., "Fungi capable of growing in strongly acid media and in concentrated copper sulfate solutions," *Journal of Bacteriology* 56 (1948), 679–681.

Oil-Degrading Bacteria

Prince, R. C., "Bioremediation of marine oil spills," *Trends in Biotechnology* 15 (1997), 158–160.

Rueter, P., *et al.*, "Anaerobic oxidation of hydrocarbons in crude oil by new types of sulphate-reducing bacteria," *Nature* 372 (1994), 455–458.

Stetter, K. O., *et al.*, "Hyperthermophilic archaea are thriving in deep North Sea and Alaskan oil reservoirs," *Nature* 365 (1993), 743–745.

Swanell, R. P., Lee, K., and McDonagh, M., "Field evaluation of marine oil spill bioremediation," *Microbiological Reviews* 60 (1996), 342–365.

Young, P., "Mouldering monuments," *New Scientist* (2.11.1996), 36–38.

Links

Scripps Institute of Oceanography: http://sio.ucsd.edu/
Woods Hole Oceanographic Institution: http://www.whoi.edu
University of Regensburg, Microbiology: http://www.biologie.uni-regensburg.de/Mikrobio/Stetter/

The Cell's Survival Kit

The Heat Shock Response

Adamowicz, M., Kelley, P. M., and Nickerson, K. W., "Detergent (sodium dodecyl sulfate) shock proteins in *Escherichia coli*," *Journal of Bacteriology* 173 (1991), 229–233.

Anfinsen, C. B., "Principles that govern the folding of protein chains," *Science* 181 (1973), 223–230.

Braig, K., *et al.*, "The crystal structure of the bacterial chaperonin GroEL at 2.8Å," *Nature* 371 (1994), 578–586.

Brissette, J. L., Russel, M., Weiner, L., and Model, P., "Phage shock protein, a stress protein of *Escherichia coli*," *Proceedings of the National Academy of Sciences of the USA* 87 (1990), 862–866.

Buchner, J., *et al.*, "GroE facilitates refolding of citrate synthase by suppressing aggregation," *Biochemistry* 30 (1991), 1586–1591.

Eckerskorn, C., and Lottspeich, F., "Combination of two-dimensional gel electrophoresis with microsequencing and amino acid composition analysis: Improvement of speed and sensitivity in protein characterization," *Electrophoresis* 11 (1990), 554–561.

Goloubinoff, P., Gatenby, A. A., and Lorimer, G. H., "GroE heat-shock proteins promote assembly of foreign prokaryotic ribulose bisphosphate carboxylase oligomers in *Escherichia coli*," *Nature* 337 (1989), 44–47.

Gross, M., Kosmowsky, I. J., Lorenz, R., Molitoris, H. P., and Jaenicke, R., "Response of bacteria and fungi to high-pressure stress as investigated by two-dimensional electrophoresis," *Electrophoresis* 15 (1994), 1559–1565.

Hemmingsen, S. M., et al., "Homologous plant and bacterial proteins chaperone oligomeric protein assembly," Nature 333 (1988), 330–334.

Hickey, E. W., and Hirshfield, I. N., "Low-pH-induced effects on patterns of protein synthesis and on internal pH in Escherichia coli and Salmonella typhimurium," Applied and Environmental Microbiology 56 (1990), 1038–1045.

Jaenicke, R., "Role of accessory proteins in protein folding," Current Opinion in Structural Biology 3 (1993), 104–112.

Jaenicke, R., Bernhardt, G., Lüdemann, H.-D., and Stetter, K. O., "Pressure-induced alterations in the protein pattern of the thermophilic archaebacterium Methanococcus thermolithotrophicus," Applied and Environmental Microbiology 54 (1988), 2375–2380.

Langer, T., et al., "Successive action of DnaK, DnaJ and GroEL along the pathway of chaperone-mediated protein folding," Nature 356 (1992), 683–689.

Martin, J., Mayhew, M., Langer, T., and Hartl, F. U., "The reaction cycle of GroEL and GroES in chaperonin-assisted protein folding," Nature 366 (1993), 228–233.

Nyström, T., and Neidhardt, F. C., "Effects of overproducing the universal stress protein UspA in Escherichia coli K-12." Journal of Bacteriology 178 (1996), 927–930.

O'Farrell, P. H., "High resolution two-dimensional electrophoresis of proteins," Journal of Biological Chemistry 250 (1975), 4007—4021.

O'Farrell, P. Z., Goodman, H. M., and O'Farrell, P. H., "High resolution two-dimensional electrophoresis of basic as well as acidic proteins," Cell 12 (1977), 1133–1142.

Saibil, H. R., et al., "ATP induces large quaternary rearrangements in a cage-like chaperonin structure," Current Biology 3 (1993), 265–273.

VanBogelen, R. A., and Neidhardt, F. C., "Ribosomes as sensors of heat and cold shock in Escherichia coli." Proceedings of the National Academy of Sciences of the USA 87 (1990), 5589–5593.

Welch, T. J., Farewell, A., Neidhardt, F. C., and Bartlett, D. H., "Stress response of Escherichia coli to elevated hydrostatic pressure," Journal of Bacteriology 175 (1993), 7170–7177.

Cold Response

Berger, F., Morellet, N., Menu, F., and Potier, P., "Cold shock and cold acclimation proteins in the psychrotrophic bacterium Arthrobacter globiformis SI55," Journal of Bacteriology 178 (1996), 2999–3007.

Chao, H., et al., "Structure–function relationship in the globular type III antifreeze protein: Identification of a cluster of surface residues required for binding to ice," Protein Science 3 (1994), 1760–1769.

Jones, P. G., VanBogelen, R. A., and Neidhardt, F. C., "Induction of proteins in response to low temperature in Escherichia coli," Journal of Bacteriology 169 (1987), 2092–2095.

Newkirk, K., *et al.*, "Solution NMR structure of the major cold shock protein (CspA) from *Escherichia coli*: Identification of a binding epitope for DNA," *Proceedings of the National Academy of Sciences of the USA* 91 (1994), 5114–5118.

Sicherl, F., and Yang, D. S. C., "Ice-binding structure and mechanism of an anti-freeze protein from winter flounder," *Nature* 375 (1995), 427–431.

Sönnichsen, F. D., Sykes, B. D., Chao, H., and Davies, P. L., "The nonhelical structure of antifreeze protein type III," *Science* 259 (1993), 1154–1157.

Storey, K. B., and Storey, J. M., "Frozen and alive," *Scientific American* 263 (December 1990), 62–67.

Amino Acid Composition

Böhm, G., and Jaenicke, R., "A structure-based model for the halophilic adaptation of dihydrofolate reductase from *Halobacterium volcanii*," *Protein Engineering* 7 (1994), 213–220.

Dym, O., Mevarech, M., and Sussman, J. L., "Structural features that stabilize halophilic malate dehydrogenase from an archaebacterium," *Science* 267 (1995), 1344–1346.

Eisenberg, H., Mevarech, M., and Zaccai, G., "Biochemical, structural and molecular genetic aspects of halophilism," *Advances in Protein Chemistry* 43 (1992), 1–62.

Frolow, F., Harel, M., Sussman, J. L., Mevarech, M., and Shoham, M., "Insights into protein adaptation to a saturated salt environment from the crystal structure of a halophilic 2Fe–2S ferredoxin," *Nature Structural Biology* 3 (1996), 452–458.

Jaenicke, R., "Protein stability and molecular adaptation to extreme conditions," *European Journal of Biochemistry* 202 (1991), 715–728.

Jaenicke, R., Schurig, H., Beaucamp, N., and Ostendorp, R., "Structure and stability of hyperstable proteins: Glycolytic enzymes from hyperthermophilic bacterium *Thermotoga maritima*," *Advances in Protein Chemistry* 48 (1996), 181–269.

Small Molecules

De Virgilio, C., Piper, P., Boller, T., and Wiemken, A., "Acquisition of thermotolerance in *Saccharomyces cerevisiae* without heat shock protein hsp 104 and in the absence of protein synthesis," *FEBS Letters* 288 (1991), 86–90.

Hensel, R., and König, H., "Thermoadaptation of methanogenic bacteria by intracellular ion concentration," *FEMS Microbiology Letters* 49 (1988), 75–79.

Hottiger, T., Boller, T., and Wiemken, A., "Correlation of trehalose content and heat resistance in yeast mutants altered in the RAS/adenylate cyclase pathway: Is trehalose a thermoprotectant?" *FEBS Letters* 255 (1989), 431–434.

Ostroy, D., *et al.*, "A new cyclopyrophosphate as a bacterial antistressor," *FEBS Letters* 298 (1992), 159–161.

Repair

Ahmad, M., and Cashmore, A. R., "*HY4* gene of *A. thaliana* encodes a protein with characteristics of a blue-light photoreceptor," *Nature* 366 (1993), 162–166.

Atkins, J. F., and Gesteland, R. F., "Genetic code: A case for *trans* translation," *Nature* 379 (1996), 769–771.

Carell, T., Epple, R., and Gramlich, V., "Synthesis of flavin-containing model compounds for DNA photolyase mediated DNA repair," *Angewandte Chemie International Edition in English* 35 (1996), 620–623.

Carroll, J. D., Daly, M. J., and Minton, D. W., "Expression of *recA* in *Deinococcus radiodurans*," *Journal of Bacteriology* 178 (1996), 130–135.

Daly, M. J., and Minton, K. W., "Resistance to radiation," *Science* 270 (1995), 1318.

Grogan, D. W., "Exchange of genetic markers at extremely high temperatures in the archaeon *Sulfolobus acidocaldarius*," *Journal of Bacteriology* 178 (1996), 3207–3211.

Hearst, J. E., "The structure of photolyase: Using photon energy for DNA repair," *Science* 268 (1995), 1858–1859.

Jentsch, S., "When proteins receive deadly messages at birth," *Science* 271 (1996), 955–956.

Keiler, K. C., Waller, P. R. H., and Sauer, R. T., "Role of a peptide tagging system in degradation of proteins synthesized from damaged messenger RNA," *Science* 271 (1996), 990–993.

Minton, K. W., "DNA repair in the extremely radioresistant bacterium *Deinococcus radiodurans*," *Molecular Microbiology* 13 (1994), 9–15.

Park, H. W., Kim, S. T., Sancar, A., and Deisenhofer, J., "Crystal structure of DNA photolyase from *Escherichia coli*," *Science* 268 (1995), 1866–1872.

Sancar, A., "Structure and function of DNA photolyase," *Biochemistry* 33 (1994), 2–9.

Sancar, A., "No 'end of history' for photolyases," *Science* 272 (1996), 48–49.

Todo, T., *et al.*, "Similarity among the *Drosophila* (6-4)photolyase, a human photolyase homolog, and the DNA photolyase-blue-light photoreceptor family," *Science* 272 (1996), 109–112.

Tu, G. F., Reid, G. E., Zhang, J. G., Moritz, R. L., and Simpson, R. J., "C-terminal extension of truncated recombinant proteins in *Escherichia coli* with a 10Sa RNA decapeptide," *Journal of Biological Chemistry* 270 (1995), 9322–9326.

Sporulation

Arigoni, F., Pogliano, K., Webb, C. D., Stragier, P., and Losick, R., "Localization of protein implicated in establishment of cell type to sites of asymmetric division," *Science* 270 (1995), 637–643.

Cano, R. J., and Borucki, M. K., "Revival and identification of bacterial spores in 25- to 40-million-year-old Dominican amber," *Science* 268 (1995), 1060–1064.

Fischman, J., "Have 25-million-year-old bacteria returned to life?" *Science* 268 (1995), 977.

Jenal, U., and Stephens, S., "Bacterial differentiation: Sizing up sporulation," *Current Biology* 6 (1996), 111–114.
Setlow, P., "Mechanisms for the prevention of damage to DNA in spores of *Bacillus* species," *Annual Reviews in Microbiology* 49 (1995), 29–54.

Symbiosis

Childress, J. J., Felbeck, H., and Somero, G. N., "Symbiosis in the deep sea," *Scientific American* 256 (May 1987), 106–112.
Barinaga, M., "Mycology: Origins of lichen fungi explored," *Science* 268 (1995), 1437.
Gargas, A., DePriest, P. T., Grube, M., and Tehler, A., "Multiple origin of lichen symbioses in fungi suggested by SSU rDNA phylogeny," *Science* 268 (1995), 1492–1495.

Links

The chaperonin home page: http://bioc02.uthscsa.edu/~seale/Chap/chap.html

Relevance of Extremes for Biotechnology and Medicine

Biotechnology

Adams, M. W. W., Perler, F. B., and Kelly, R. M., "Extremozymes: Expanding the limits of biocatalysis," *Bio/technology* 13 (1995), 662–668.
Govardhan, C. P., and Margolin, A. L., "Extremozymes for industry—From nature and by design," *Chemistry & Industry* (4.9.1995), 689–693.
Franks, F., "Long-term stabilization of biologicals," *Bio/technology* 12 (1994), 253–256.
Hayashi, R., "Utilization of pressure in addition to temperature in food science and technology," in *High Pressure and Biotechnology*, C. Balny, R. Hayashi, K. Heremans, and P. Masson, eds. (London: John Libbey, 1992), pp. 185–193.
Hoover, D. G., Metrick, C., Papineau, A. M., Farkas, D. F., and Knorr, D., "Biological effects of high hydrostatic pressure on food microorganisms," *Food Technology* 43 (1989), 99–107.
Johnston, D. E., "High pressure—A new dimension to food processing," *Chemistry & Industry* (4.7.1994), 499–501.
Mertens, B., and Knorr, D., "Developments of nonthermal processes for food preservation," *Food Technology* 46 (1992), 124–133.
Mozhaev, V. V., Heremans, K., Frank, J., Masson, P., and Balny, C., "Exploiting the effects of high hydrostatic pressure in biotechnological applications," *Trends in Biotechnology* 12 (1994), 493–501.

Mullis, K. B., "The unusual origin of PCR" *Scientific American* 262 (April 1990), 36.
Woodward, J. W., *et al.*, "*In vitro* hydrogen production by glucose dehydrogenase and hydrogenase," *Nature Biotechnology* 14 (1996), 872–874.

Bacteriorhodopsin Applications

Birge, R. R., "Protein-based computers," *Scientific American* 272 (March 1995), 66–71.
Birge, R. R., "Protein-based three-dimensional memory," *American Scientist* 82 (1994), 348–355.
Bräuchle, C., Hampp, N., and Oesterheld, D., "Optical applications of bacterio-rhodopsin and its mutated variants," *Advanced Materials* 3 (1991), 420–428.
Hampp, N., Bräuchle, C., and Oesterheld, D., "Bacteriorhodopsin wildtype and variant aspartate-96→asparagine as reversible holographic media," *Biophysical Journal* 58 (1990), 83–93.
Koyama, K., Yamaguchi, N., and Miyasaka, T., "Antibody-mediated bacterio-rhodopsin orientation for molecular device architectures," *Science* 265 (1994), 762–765.
Oesterheld, D., Bräuchle, C., and Hampp, N., "Bacteriorhodopsin: A biological material for information processing," *Quarterly Reviews of Biophysics* 24 (1991), 425–478.
Stoeckenius, W., "The purple membrane of salt-loving bacteria," *Scientific American* 234 (June 1976), 38–46.
Vsevolodov, N. N., and Dyukova, T. V., "Retinal-protein complexes as optoelec-tronic components," *Trends in Biotechnology* 12 (1994), 81–88.

Medicine

Blaser, M. J., "The bacteria behind ulcers," *Scientific American* 274 (February 1996), 92–97.
Edgington, S. M., "Therapeutic applications of heat shock proteins," *Bio/technology* 13 (1995), 1442–1444.
Jindal, S., "Heat shock proteins: Applications in health and disease," *Trends in Biotechnology* 14 (1996), 17–20.

Extremists and the Tree of Life

Origin of Life

Allègre, C. J., and Schneider, S. H., "The evolution of the earth," *Scientific American* 271 (October 1994), 44–51.

Balter, M., "Looking for clues to the mystery of life on earth," *Science* 273 (1996), 870–872.

Cairns-Smith, A., *Seven Clues to the Origin of Life—A Scientific Detective Story* (Cambridge: Cambridge University Press, 1985).

Chyba, C., and Sagan, C., "Endogenous production, exogenous delivery and impact-shock synthesis of organic molecules: An inventory for the origins of life," *Nature* 355 (1992), 125–132.

Ferris, J. P., Hill, A. R., Jr., Liu, R., and Orgel, L. E., "Synthesis of long prebiotic oligomers on mineral surfaces," *Nature* 381 (1996), 59–61.

Hayes, J. M., "The earliest memories of life on Earth," *Nature* 384 (1996), 21–22.

Keller, M., Blöchl, E., Wächtershäuser, G., and Stetter, K. O., "Formation of amide bonds without a condensation agent and implications for origin of life," *Nature* 368 (1994), 836–838.

Lazcano, A., and Miller, S. L., "The origin and early evolution of life: Prebiotic chemistry, the pre-RNA world, and time," *Cell* 85 (1996), 793–798.

Mojzsis, S. J., Arrhenius, G., McKeegan, K. D., Harrison, G. M., Nutman, A. P., and Friend, C. R. L., "Evidence for life on Earth before 3,800 million years ago," *Nature* 384 (1996), 55–59.

Orgel, L. E., "The origin of life on the earth," *Scientific American* 271 (October 1994), 52–61.

Pace, N. R., "Origin of life—Facing up to the physical setting," *Cell* 65 (1991), 531–533.

Piccirilli, J. A., "Origin of life: RNA seeks its maker," *Nature* 376 (1995), 548–549.

Schwartz, A. W., "Did minerals perform prebiotic combinatorial chemistry?" *Chemistry & Biology* 3 (1996), 515–518.

Von Kiedrowski, G., "Origins of life: Primordial soup or crepes?" *Nature* 381 (1996), 20–21.

RNA-World

Cate, J. H., *et al.*, "Crystal structure of a group I ribozyme domain: Principles of RNA packing," *Science* 273 (1996), 1678–1685.

Cate, J. H., *et al.*, "RNA tertiary structure mediation by adenosine platforms," *Science* 273 (1996), 1696–1699.

Cech, T. R., "RNA as an enzyme," *Scientific American* 255 (November 1986), 76–84.

Ekland, E. H., and Bartel, D. P., "RNA-catalyzed RNA polymerization using nucleoside triphosphates," *Nature* 382 (1996), 373–376.

Ilangasekare, M., Sanchez, G., Nickles, G., and Yarus, M., "Aminoacyl-RNA synthesis catalyzed by an RNA," *Science* 267 (1995), 643.

Jones, J. T., Lee, S. W., and Sullenger, B. A., "Tagging ribozyme reaction sites to follow *trans*-splicing in mammalian cells," *Nature Medicine* 2 (1996), 643–648.

Joyce, G. F., "Ribozymes: Building the RNA world," *Current Biology* 6 (1996), 965–967.

Lohse, P. A., and Szostak, J. W., "Ribozyme-catalysed amino-acid transfer reactions," *Nature* 381 (1996), 442–444.
Michel, F., and Westhof, E., "Visualizing the logic behind RNA self-assembly," *Science* 273 (1996), 1676–1677.
Noller, H. F., Hoffarth, V., and Zimniak, L., "Unusual resistance of peptidyl transferase to protein extraction procedures," *Science* 256 (1992), 1416–1419.
Piccirilli, J. A., McConnell, T. S., Zaug, A. J., Noller, H. F., and Cech, T. R., "Aminoacyl esterase activity of the *Tetrahymena* ribozyme," *Science* 256 (1992), 1420–1423.
Scott, W. G., and Klug, A., "Ribozymes: Structure and mechanism in RNA catalysis," *Trends in Biochemical Sciences* 21 (1996), 220–224.

Archaebacteria

Bult, C. J., *et al.*, "Complete genome sequence of the methanogenic archaeon, *Methanococcus jannaschii*," *Science* 273 (1996), 1058–1073.
Gray, M. W., "Genomics: The third form of life," *Nature* 383 (1996), 299–300.
Morell, V., "Life's last domain," *Science* 273 (1996), 1043–1045.
Woese, C. R., "Archaebacteria," *Scientific American* 244 (June 1981), 94–106.

Hot Origins

Forterre, P., "A hot topic: The origin of hyperthermophiles," *Cell* 85 (1996), 789–792.
Nisbet, E. G., and Fowler, C. M. R., "Early life: Some liked it hot," *Nature* 382 (1996), 404–405.

Gaia

Kump, L. R., "Gaia: The physiology of the planet," *Nature* 381 (1996), 111–112.
Lovelock, J. E., *Gaia: A New Look at Life on Earth* (Oxford: Oxford University Press, 1979).
Lovelock, J. E., and Kump, L. R., "Failure of climate regulation in a geophysiological model," *Nature* 369 (1994), 732–734.

Links

The Institute for Genomic Research: http://www.tigr.org./

Life beyond Earth

Life on Other Planets

Chyba, C. F., "Exobiology: Life beyond Mars," *Nature* 382 (1996), 576–577.

Grady, M., Wright, I., and Pillinger, C., "Opening a Martian can of worms," *Nature* 382 (1996), 575–576.

Kerr, R. A., "Ancient life on Mars?" *Science* 273 (1996), 864–866.

McKay, D. S., *et al.*, "Search for past life on Mars: Possible relic biogenic activity in Martian meteorite ALH84001," *Science* 273 (1996), 924–930.

Sagan, C., "The search for extraterrestrial life," *Scientific American* 271 (October 1994), 70–77.

Sagan, C., Thompson, W. R., Carlson, R., Gurnett, D., and Hord, C., "A search for life on Earth from the *Galileo* spacecraft," *Nature* 365 (1993), 715–721.

Extrasolar Planets

Angel, J. R. P., and Woolf, N. J., "Searching for life on other planets," *Scientific American* 274 (April 1996), 46–52.

Beckwith, S. V. W., and Sargent, A. I., "Circumstellar disks and the search for neighbouring planetary systems," *Nature* 383 (1996), 139–144.

Parsons, P., "Exobiology: Dusting off panspermia," *Nature* 383 (1996), 221–222.

Walker, G. A. H., "Extrasolar planets: A solar system next door," *Nature* 382 (1996), 23–24.

Williams, D. M., Kasting, J. F., and Wade, R. A., "Habitable moons around extra-solar giant planets," *Nature* 385 (1997), 234–236.

Links

NASA origins program: http://origins.stsci.edu/
Mars Exploration Program: http://mpfwww.jpl.nasa.gov/
Mars Global Surveyor: http://www.jpl.nasa.gov/mgs/
Kepler (search for habitable planets): http://www.kepler.arc.nasa.gov/
Extrasolar Planets Encyclopedia: http://www.obspm.fr/planets

Figure Acknowledgments

Permission to reproduce the following figures is gratefully acknowledged:

Front cover: Photographer/Science Photo Library, 112 Westbourne Grove, London, England, W2 5RU.

Figure 4, Chapter 2: By permission from Patricia J. Wynne, artist.

Figure 5, Chapter 2: Dr. Reinhard Rachel, Universität Regensburg, Lehrstuhl für Biologie VII, Universitatsstr. 31, 93040 Regensburg, Germany.

Figure 11, Chapter 2: HOCHTIEF UMWELT GmbH, Huyssenallee 86–88, 45128 Essen, Germany.

Figure 2, Chapter 3: Professor Udo Heinemann, Max-Delbruck-Centrum für Molekulare Medizin, Robert-Rossle-Str. 10, 13122 Berlin, Germany. Figure of CspA from *Proceedings of the National Academy of Sciences of the USA*, 91 (1994), 5119–5123.

Figure 6, Chapter 3: By permission from Patricia J. Wynne, artist.

Afterword
Latest News from the Edge

The manuscript for the hardback edition of Life on the Edge was finished in September, 1997—but life has moved on since then, and so has scientific exploration of life under extreme conditions. New extreme habitats have been discovered, exploration of potential habitats such as Lake Vostok, Europa, and Titan has progressed, while the Mars program has suffered some embarrassing drawbacks. New connections between environmental stress and evolution have emerged, while shocking revelations have been made about not too distant climate changes on our planet.

Where statements made in the original text needed updating, new endnotes have been added to that effect. Some important new issues that have come up are described here in more detail, and arranged in the order in which the issues would have appeared in the contents of the book.

In Chapter 3, I explained how some heat shock proteins serve as folding helpers both in stress situations and in everyday life. From the demonstration of how this works in the test tube, the field has now moved on to address a new question.

Who Needs a Chaperone?
Customer Files of Bacterial GroEL Revealed

It has been known since the early days of chaperone research that GroEL can be a folding helper for many proteins (both from its native *Escherichia coli* and from other organisms), but not for all. However, it was far from clear what distinguishes its substrates from those proteins upon which it has no effect. The completion of the *E. coli* genome sequence in 1997 offered the opportunity to investigate the real biological significance of this chaperone in its native organism and to answer the question of which proteins rely on its services. Knowing the genome sequence, one could essentially separate the proteins bound to GroEL on a 2D gel (see box on page 63), determine the beginning of the sequence for each spot, and look it up in the genome database.

For this project, the group of Ulrich Hartl (now at the Max Planck Institute for Biochemistry) collaborated with Christoph Eckerskorn and Friedrich Lottspeich, who pioneered such methods of analysis from 2D gels. In order to be able to distinguish between proteins chaperoned immediately following their synthesis and those coming back to the GroEL workshop for service and repair, they had to introduce a time axis. Therefore, they fed the bacteria a radioactive nutrient for a very brief time interval (the pulse). The radioactivity would turn up only in those proteins that were being synthesized at the moment of the pulse. By varying the time intervals between the pulse and the harvesting of the cells, they were able to look at the populations of molecules bound to GroEL at specific times after synthesis.

Analysis following the pulse revealed 250–300 kinds of proteins that obviously required the help of GroEL directly after their synthesis. This is around a tenth of the number of proteins normally present in the *E. coli* cytoplasm. Most of the substrates were in the middle size range, between 20 and 60 kilodaltons. This makes sense, as molecules in this range are big enough to exhibit complex folding patterns, but not too big to fit into the cavity of GroEL.

When the time interval is increased, the number of proteins declines rapidly. The majority seem to have left GroEL within a few minutes. At time intervals between ten minutes and two hours, however, the researchers observed a smaller but persistent group of proteins that obviously come back to GroEL for repeated servicing operations. In the time from one cell division to the next (around one hour under normal conditions), these proteins return about four times on average, taking up roughly 30% of the GroEL capacity available. Under heat shock conditions, however, the fraction needed for service and repair can be up to 60 %.

By partially sequencing the minute amount of (very pure) protein contained in each spot found in 2D gels, the researchers could unambiguously identify 52 of the service customers. As in earlier comparisons of smaller numbers of substrate proteins, no shared amino acid sequence could be found which might aid recognition. Apart from the size preference mentioned above, the most significant property that these proteins had in common is a high abundance of domains containing both alpha helices and beta sheets (as opposed to folding units made up of only one of these secondary structure types). This finding is quite plausible, as it is known from folding studies that such mixed domains can be difficult to fold.

Heat shock proteins can have other functions apart from being chaperones. One of them, the previously little-known Hsp90, emerged as a crucial switch in development, a finding that may have far-reaching consequences with respect to the connections between environmental stress and evolution.

A Heat Shock Protein Accounts for Bouncy Evolution

In a dispute dubbed "the Darwin wars" by some, the authors Richard Dawkins and Stephen Jay Gould famously disagree about the way evolution progresses. While Dawkins holds up the view that the steady stream of random mutations should make evolution flow smoothly, Gould argues that unidentified processes apparently made evolution behave more irregularly. According to his view, long phases with little change could alternate with rapid outbursts of inventiveness. Until very recently, there was no evidence for a mechanism that could have accounted for this "punctuated equilibrium" scenario favored by Gould. In 1998, however, an utterly surprising discovery involving a heat shock protein from the fruit fly *Drosophila* provided a neat biochemical link between environmental stress and the speed of evolutionary change.

The protein in question, Hsp90, was among those heat shock proteins whose functions proved to be less clearcut than that of the classical GroEL described in Chapter 3. Like GroEL, however, and unlike some other stress proteins, it is already very strongly present in the cell in the absence of stress. In heat shock conditions, Hsp90 acts as a chaperone, keeping unfolded proteins out of trouble. But at normal conditions, its role is apparently associated with signalling processes within the cell. Certain molecular switches meant to recognize a given hormone molecule and then do something about it (the steroid hormone receptors) were found to be fully active only when they were associated with an assembly of several other protein molecules including two copies of Hsp90. This, together with the presence of other chaperones in this complex, suggests that Hsp90 might help the steroid receptors maintain the correctly folded state required for their task. This finding was still compatible with the current knowledge about heat shock proteins and chaperones.

Suzanne Rutherford and Susan Lindquist investigated fruit flies that had lost one of the two copies of their Hsp90 gene. (Losing both copies is lethal, but flies with one copy are viable and fertile and can thus be used for cross-breeding experiments.) The initial observation made in Lindquist's lab at the University of Chicago was that populations of flies with only one copy of the Hsp90 gene came up with malformations (e.g., alterations in the structures of eyes or wings) much more often than normal fly populations.

Investigating this phenomenon more closely, the researchers made four observations that suggested to them that the function of Hsp90 must be related to these malformations:

- Mutants of different origin but with similar mutations in Hsp90 had offspring with similar malformations.
- Double mutants, in which both copies of the gene were altered in different, moderate (non-lethal) ways, had even more severe malformations.
- Feeding a specific Hsp90 inhibitor to flies with two healthy copies of the gene resulted in offspring with similar malformations.
- Breeding healthy flies at very high or very low temperatures equally resulted in malformations.

All of these observations suggest that whenever Hsp90 is weakened in any way (by genetic mutation, by the presence of an inhibitor, or by being summoned to assist the stress response), malformations are more likely to appear. This could mean that Hsp90 had an as yet unidentified role in lowering the error rate in DNA copying. Further cross-breeding experiments suggested, however, that the genetic mutations responsible for the malformations were already present as silent variants in the healthy parents and ancestors of the malformed flies. Hsp90 appears to have stopped these mutations from affecting the appearance (the phenotype) of the fly.

According to this interpretation, Hsp90 allows a fly population to "store" genetic variability in a silent way over a number of generations, and to activate it only in times of environmental stress. Although the malformations are disadvantageous for many of the affected individuals, the strategy makes sense for the population as a whole. If environmental conditions change rapidly in a way that requires them to use their stress proteins permanently, there is no time left to hope for gradual evolutionary change. When the extinction of the whole population or even species is at stake, production of offspring with a wide variety of genetic properties improves the chances that there will be some descendents that will be able to cope with the new conditions. Thus, Hsp90 would serve as a buffer that can save up genetic variability for when it may be needed. From the outside, this would look just like a punctuated equilibrium. Nothing seems to change for a long time, then drastic changes occur very rapidly.

Apart from stress proteins, there are also some smaller molecules that can protect cells from environmental stress. Among these, the sugar trehalose, introduced on page 78, was found to be involved in a feat of record-breaking stress resistance.

Tardigrades Can Take the Pressure

In 1998, Kunihiro Seki and Masato Toyoshima (at Kanagawa University, Japan) reported the unusual results they obtained studying the pressure resistance of two species of tardigrades. These microscopically small animals reminiscent of downsized bears are at most a half-millimeter long, live in water droplets suspended in moss and lichens, and can be found on all continents. They have at least two major emergency routines. If their habitat is flooded and there is a risk of oxygen shortage, they inflate to a balloon-like passive state that can float around on the water for days. If, however, the threat comes from a lack of water, they shrink to form the so-called tun state (because it looks like a barrel), which could be described as the animal equivalent of a spore. Tardigrades have been resuscitated by rehydrating moss samples after up to 100 years of storage on museum shelves, which proves the quite remarkable long-term stability of this state.

It was this tun state that the Japanese researchers used in their high pressure studies. As the presence of water would have converted the animals back to the active state, the researchers suspended the tuns in a perfluorocarbon solvent before they applied pressures of up to 6,000 atmospheres (more than five-fold the pressure found in the deepest trenches of the oceans). While active tardigrade populations in water are killed off by 2000 atmospheres (already a remarkably high threshold for an animal), the tun state allowed 95% of the individuals of one species and 80% of another to survive the maximal pressure of 6,000 atmospheres.

This observation is unprecedented for any animal species. Only some bacterial spores and lichens could hope to compete. Still, tardigrade experts may have been only mildly surprised, as they knew already that the tuns could be revived after freezing in liquid helium—they are frost resistant down to 0.5 Kelvin. Detailed mechanistic explanations for these record-breaking achievements are not yet available. One thing that is known for sure is that the tuns contain high concentrations of the sugar trehalose, which is known to improve the stress resistance of baker's yeast (see pages 78-79). The phenomenal shelf life of the tuns has aroused the interest of researchers in medical technology. Some are trying to copy the tardigrades' recipe to achieve similar long-term stability for human organs to be used in transplantation.

The discoveries of more and more mechanisms by which living organisms can survive extreme environmental conditions coincides with the realization that there have been times in Earth's not too distant history when these qualities were in high demand.

Escape From Frozen Hell (Why we should be glad that Earth turned into a snowball not so very long ago)

Some very weird things happened on our planet some 700 to 600 million years ago, in a time geologists call the Neoproterozoic era. Apparently, glaciers were pushing gravel around in places that were near the equator and not very high above sea level. The distribution of the isotopes (different kinds of atoms) of carbon found in minerals dating from this time suggests that life (which prefers one kind of carbon over the other, and thus shifts the balance) must have been very nearly absent.

In August, 1998, Harvard geologist Paul Hoffman, together with three geochemist colleagues, proposed a daring scenario that neatly explained the weird observations, but also fueled controversy. According to their "Snowball Earth" claim, our planet was completely frozen not only once, but twice, during the Neoproterozoic era. At that time, the Sun was somewhat fainter than today. The break-up of a supercontinent may have exposed large amounts of rock surface, which, by weathering, withdrew the greenhouse gas carbon dioxide from the atmosphere. As these new continents were located near the equator, they were not immediately covered by the advancing glaciation, so their carbon dioxide sucking effect persisted.

The trouble with freezing planets is that the more ice you have on the surface, the more energy from the Sun gets reflected back into space, so the planet gets even colder, and more widely covered in ice . . . until it is a perfect snowball. The Russian climatologist Mikhail Budyko had shown in the 1960s that such runaway glaciation (technically described as a positive feedback loop) could happen in principle. But as we are here to witness that Earth is not a snowball right now, Hoffman and colleagues had to find a way of turning the runaway glaciation around.

Caltech geologist Joe Kirschvink first proposed in 1986 how the snowball defrosted, but the idea didn't find much resonance at the time. It relies on the observation that volcanoes spit the greenhouse gas carbon dioxide into the atmosphere. Nowadays, this effect is balanced out by other processes that remove carbon dioxide—mainly photosynthesis, weathering of minerals, and washing out by rain and snow. On snowball Earth, however, the water was locked in the ice and underneath it, so there was no washout, or any exposed rocks or photosynthesis activities, either. So the greenhouse gas that the volcanoes exhaled built up in the atmosphere, until, after a few million years, enough global warming capacity had built up to melt the ice. In a massive overshoot effect, the planet then turned into a global sauna, until most of the carbon dioxide was washed out into the oceans and from there deposited as carbonate minerals—another part of the previously unexplained geological weirdness dating from this era.

Nobody denies that runaway glaciation might have happened. Much of the discussion triggered by the snowball scenario was concerned with biological than with geological implications. How could the biosphere survive such a dramatic threat? In fact, recent research into the most extreme conditions that life can tolerate on our planet suggests it is quite possible that life survived the two freeze/fry cycles. Earlier generations of geologists were aware of the evidence of glaciation near the equator, but they dismissed the possibility of runaway glaciation as incompatible with the fact that the biosphere survived. But now that we know more about life under extreme conditions, snowball Earth may be an idea whose time has come.

One obvious survival niche would have been given by small leftovers of sea water or molten pools near the glaciers. A computer simulation of the snowball phase published by William T. Hyde from Texas A&M University in May, 2000 suggests that a belt of open sea water may have continued to exist near the equator. Algae could have survived there. Lichens and bacterial spores could have survived even in the ice for extended periods of time, although we cannot be sure that this strategy would have worked for millions of years. As biologists now know that certain organisms can thrive at temperatures above boiling or below freezing, they would not necessarily assume that a period of extreme temperature changes would extinguish life on Earth.

The episode did, however, reduce the activity of life to a very small fraction of what it was before—that can be deduced from the unusual records of carbon isotopes. The number of species certainly fell dramatically. Those species that did survive probably were forced to develop or reactivate efficient mechanisms to deal with temperature changes, including those described in Chapter 3.

When moderate climate conditions returned, the few, well-seasoned species that survived found a whole planet with many ecological niches to colonize from scratch. Such conditions encourage the evolution of a large number of new species. After all, the mass extinction that wiped out the dinosaurs (along with many other less popular species) provided a unique opportunity for a hitherto marginal group of animals now known as mammals. Similarly, the tabula rasa situation after the last snowball episode would have encouraged new species to emerge on a big scale.

Which is exactly what paleontologists find in the fossil record. Shortly after the big freeze, the strange looking Ediacara fauna shows up, evolution's first attempt at building multicellular animals. This is probably not a coincidence, any more than the fact that fewer than fifty million years after the last thawing of snowball Earth, the Cambrian explosion took place, in which all of the animal body plans around today were developed. Plants and animals established a much more complex biosphere including negative feedback loops (damping climate change rather than amplifying it) that, together with a slightly brighter sun, has spared us any further glaciation catastrophes.

Until very recently, scientists thought that if our planet had frozen over completely in the Neoproterozoic, we wouldn't be here to talk about it. In the light of what we have learned about life under extreme conditions, however, it appears that the contrary may be true: If Earth had not turned into a snowball, it might still be dominated by microbes, and we wouldn't be here to appreciate the development of more complex life forms.

Index